The Best
WRITING on
MATHEMATICS

2020

The **BEST**
WRITING on
MATHEMATICS

2020

Mircea Pitici, Editor

PRINCETON UNIVERSITY PRESS
PRINCETON AND OXFORD

Copyright © 2020 by Princeton University Press

Princeton University Press is committed to the protection of
copyright and the intellectual property our authors entrust
to us. Copyright promotes the progress and integrity of
knowledge. Thank you for supporting free speech and the
global exchange of ideas by purchasing an authorized edition
of this book. If you wish to reproduce or distribute any part of
it in any form, please obtain permission.

Requests for permission to reproduce material from this work
should be sent to permissions@press.princeton.edu

Published by Princeton University Press
41 William Street, Princeton, New Jersey 08540
6 Oxford Street, Woodstock, Oxfordshire OX20 1TR

press.princeton.edu

All Rights Reserved

ISBN 9780691207575
ISBN (pbk.) 9780691207568
ISBN (e-book) 9780691213651

British Library Cataloging-in-Publication Data is available

Editorial: Susannah Shoemaker and Kristen Hop
Production Editorial: Nathan Carr
Text Design: Carmina Alvarez
Jacket/Cover Design: Chris Ferrante
Production: Jacquie Poirier
Publicity: Matthew Taylor and Katie Lewis
Copyeditor: Paula Bérard

This book has been composed in Perpetua

Printed on acid-free paper. ∞

Printed in the United States of America

1 3 5 7 9 10 8 6 4 2

*to the front-end management of
the Ithaca Wegmans supermarket,
for making a difference to me
and to my family——twice*

Contents

Color illustration insert follows page 128

Introduction

MIRCEA PITICI

This anthology is the 11th in the annual series of *The Best Writing on Mathematics*. It contains pieces originally published in late 2018 and throughout 2019 in various venues, including specialized print and online magazines, research journals, newspapers, books, and collections of conference proceedings. The volume should be considered by the readers in conjunction with the other ten previously published in the series.

Overview of the Volume

In a piece eerily reminding us of the current coronavirus health crisis, Steven Strogatz recounts the little-known contribution of differential equations to virology during the HIV crisis and makes the case for considering calculus among the heroes of modern life.

Peter Denning and Ted Lewis examine the genealogy, the progress, and the limitations of complexity theory—a set of principles developed by mathematicians and physicists who attempt to tame the uncertainty of social and natural processes.

In yet another example of fusion between ideas from mathematics and physics, Bruce Boghosian describes how a series of simulations carried out to model the long-term outcome of economic interactions based on free-market exchanges inexorably leads to extreme inequality and to the oligarchical concentration of wealth.

Stan Wagon points out the harmonic-average intricacies, the practical paradoxes, and the policy implications that result from using the miles-per-gallon measure for the fuel economy of hybrid cars.

Jørgen Veisdal details some of the comparative reasoning supposed to take place in majoritarian democracies—resulting in electoral strategies that lead candidates toward the center of the political spectrum.

In an autobiographical piece, John Baez narrates the convoluted professional path that took him, over many years, closer and closer to algebraic geometry—a branch of mathematics that offers insights into the relationship between the classical mechanics and quantum physics.

Erica Klarreich explains how Hao Huang used the combinatorics of cube nodes to give a succinct proof to a long-standing computer science conjecture that remained open for several decades, despite many repeated attempts to settle it.

A graph-based explanation, combined with a stereographic projection, also helped Richard Montgomery solve one of the questions posed by the dynamical system formed by three masses moving under the reciprocal influences of their gravitational pulls, also known as the three-body problem.

Chris King, who created valuable online resources freely available to everyone, describes the algebraic iterations that lead to families of fractal-like, visually stunning geometric configurations and stand at the confluence of multiple research areas in mathematics.

In the next contribution to our volume, Jim Henle presents several paper-and-pencil games selected from the vast collection invented by Sid Sackson.

Dave Linkletter breaks the classic Rubik's cube apart and, using the mechanics of the cube's skeleton, counts for us the total number of possible configurations; then he reviews a collection of mathematical questions posed by the toy—some answered and some still open.

Colin Adams introduces with examples, defines, and discusses several important properties of the hyperbolic 3-manifold, a geometric notion both common to our physical environment and difficult to understand in its full generality.

In a similar geometric vein, with yet more examples, physical models, and definitions, followed by applications, Boris Odehnal presents an overview of higher dimensional geometries.

With linguistic flourishes recalling Fermat's cryptic style, James Propp traces the history of two apparently disconnected results in the theory of numbers—which, surprisingly, turned out to be strongly related—and tells us how an amateur mathematician used the parallelism to prove one of them.

Patrick Honner works out in several different ways a simple multiplication example to compare the computational efforts required by the algorithms used in each case and to illustrate the significant benefits that result when the most efficient method is scaled up to multiply big numbers.

Ben Orlin combines his drawing and teaching talents to prove that ignorance of widely known mathematics can be both hilariously ridiculous and academically rewarding!

Donald Teets's piece is entirely concerned with the young Karl Friedrich Gauss's contribution to the history of the Christian calendar.

Paul Thagard proposes five conjectures (and many more puzzling questions) on the working of mathematics in mind and society and formulates an eclectic metaphysics that affirms both realistic and fictional qualities for mathematics.

Mark Colyvan asserts that explanation in mathematics—unlike explanation in sciences and in general—is neither causal nor deductive; instead, depending on the context, mathematical explanation provides either local insights that connect similar mathematical situations or global answers that arise from non-mathematical phenomena.

Gerry Hahn, Necip Doganaksoy, and Bill Meeker call (as they have done over a long period of time) for improving statistical inquiry and analysis by using new tools—such as tolerance and prediction intervals, as well as a refined analysis of the role of sample size in experiments.

More Writings on Mathematics

Readers of this series of anthologies know that in each volume I offer many other reading suggestions from the recent literature on mathematics: book titles in the Introduction and articles in the section on Notable Writings, toward the end of the volume. As a matter of principle, I never included in these lists materials I have not seen; thus, my ability to keep up with the literature has been considerable affected by the health crisis that closed university campuses and libraries during the spring of 2020. I thank the authors and the publishers who sent me books over the last year; complete references are at the end of this introduction.

To start my book recommendation list, special mention deserves—for exceptional illustrations and insightful contributions—the collective volume published by the Bodleian Library with the title *Thinking 3D*, edited by Daryl Green and Laura Moretti. Also—for visual aspect, inspired humor, and teaching insights—Ben Orlin's books *Math with Bad Drawings* and *Change Is the Only Constant*.

Excellent expository introductions to specific topics are Julian Havil's *Curves for the Mathematically Curious*, Steven Strogatz's *Infinite Powers*, and (slightly more technical) David Feldman's *Chaos and Dynamical Systems*.

In applied mathematics and connections to other domains, we have *The Mathematics of Politics* by Arthur Robinson and Daniel Ullman, *Modelling Nature* by Edward and Michael Gillman, *Data Analysis for the Social Sciences* by Douglas Bors, *Islands of Order* by Stephen Lansing and Murray P. Cox, *Producers, Consumers, and Partial Equilibrium* by David Mandy, and *Ranking: The Unwritten Rules of the Social Game We All Play* by Péter Érdi. Featuring mathematics in astronomy are *Finding Our Place in the Solar System* by Todd Timberlake and Paul Wallace and *Our Universe* by Jo Dunkley; on mathematics in military affairs, *The (Real) Revolution in Military Affairs* by Andrei Martyanov. Two expository statistics books are *The Art of Statistics* by David Spiegelhalter and *Statistics in Social Work* by Amy Batchelor; and an excursion into computer science is *Computational Thinking* by Peter Denning and Matti Tedre.

Interdisciplinary with historical elements but also with ramifications in contemporary affairs are *Proof! How the World Became Geometrical* by Amir Alexander and *How Charts Lie* by Alberto Cairo. Several books last year were dedicated to the increasing role of algorithms in daily social affairs, including *Algorithmic Regulation* edited by Karen Yeung and Martin Lodge, *The Ethical Algorithm* by Michael Kearns and Aaron Roth, and *The Information Manifold* by Antonio Badia.

A few recent books on the history of mathematics are *Power in Numbers* by Talithia Williams, *Bernard Bolzano* by Paul Rusnock and Jan Šebestík, and *David Hume on Miracles, Evidence, and Probability* by William Vanderburgh. Also historical, with strong reciprocal influences between mathematics and the cultural, social, and linguistics contexts are *Disharmony of the Spheres* by Jennifer Nelson, *Republic of Numbers* by David Lindsay Roberts, and *Roads to Reference* by Mario Gómez-Torrente. In logic and philosophy of mathematics is *Reflections on the Foundations of Mathematics* edited by Stefania Centrone, Deborah Kant, and Deniz Sarikaya. Mathematical notions in practical philosophy appear in *Measurement and Meaning* by Ferenc Csatári and in *Conscious Action Theory* by Wolfgang Baer.

◎╅◎

I hope that you, the reader, will enjoy reading this anthology at least as much as I enjoyed working on it. I encourage you to send comments, suggestions, and materials I might consider for (or mention in) future volumes to Mircea Pitici, P.O. Box 4671, Ithaca, NY 14852; or electronic correspondence to mip7@cornell.edu.

Books Mentioned

Alexander, Amir. *Proof! How the World Became Geometrical*. New York: Farrar, Straus, and Giroux, 2019.

Badia, Antonio. *The Information Manifold: Why Computers Can't Solve Algorithmic Bias and Fake News*. Cambridge, MA: MIT Press, 2019.

Baer, Wolfgang. *Conscious Action Theory: An Introduction to the Event-Oriented World View*. Abingdon, U.K.: Routledge, 2019.

Batchelor, Amy. *Statistics in Social Work: An Introduction to Practical Applications*. New York: Columbia University Press, 2019.

Bors, Douglas. *Data Analysis for the Social Sciences: Integrating Theory and Practice*. Thousand Oaks, CA: SAGE Publications, 2018.

Cairo, Alberto. *How Charts Lie: Getting Smarter about Visual Information*. New York: W. W. Norton, 2019.

Centrone, Stefania, Deborah Kant, and Deniz Sarikaya. (Eds.) *Reflections on the Foundations of Mathematics: Univalent Foundations, Set Theory and General Thoughts*. Cham, Switzerland: Springer Nature, 2019.

Csatári, Ferenc. *Measurement and Meaning*. Lanham, MD: Rowman & Littlefield, 2019.

Denning, Peter J., and Matti Tedre. *Computational Thinking*. Cambridge, MA: MIT Press, 2019.

Dunkley, Jo. *Our Universe: An Astronomer's Guide*. Cambridge, MA: Harvard University Press, 2019.

Érdi, Péter. *Ranking: The Unwritten Rules of the Social Game We All Play*. New York: Oxford University Press, 2019.

Feldman, David P. *Chaos and Dynamical Systems*. Princeton, NJ: Princeton University Press, 2019.

Gillman, Edward, and Michael Gillman. *Modelling Nature: An Introduction to Mathematical Modelling of Natural Systems*. Wallingford, U.K.: CAB International, 2019.

Gómez-Torrente, Mario. *Roads to Reference: An Essay on Reference Fixing in Natural Language*. Oxford, U.K.: Oxford University Press, 2019.

Green, Daryl, and Laura Moretti. (Eds.) *Thinking 3D: Books, Images and Ideas from Leonardo to the Present*. Oxford, U.K.: Bodleian Library, 2019.

Havil, Julian. *Curves for the Mathematically Curious: An Anthology of the Unpredictable, Historical, Beautiful, and Romantic*. Princeton, NJ: Princeton University Press, 2019.

Kearns, Michael, and Aaron Roth. *The Ethical Algorithm: The Science of Socially Aware Algorithm Design*. New York: Oxford University Press, 2019.

Lansing, J. Stephen, and Murray P. Cox. *Islands of Order: A Guide to Complexity Modeling for the Social Sciences*. Princeton, NJ: Princeton University Press, 2019.

Mandy, David M. *Producers, Consumers, and Partial Equilibrium.* London: Elsevier, 2017.

Martyanov, Andrei. *The (Real) Revolution in Military Affairs.* Atlanta: Clarity Press, 2019.

Nelson, Jennifer. *Disharmony of the Spheres: The Europe of Holbein's Ambassadors.* University Park, PA: Penn State University Press, 2019.

Orlin, Ben. *Change Is the Only Constant: The Wisdom of Calculus in a Madcap World.* New York: Black Dog & Leventhal, 2019.

Orlin, Ben. *Math with Bad Drawings: Illuminating the Ideas That Shape Our Reality.* New York: Black Dog & Leventhal, 2018.

Roberts, David Lindsay. *Republic of Numbers: Unexpected Stories of Mathematical Americans through History.* Baltimore, MD: Johns Hopkins University Press, 2019.

Robinson, E. Arthur, and Daniel H. Ullman. *The Mathematics of Politics,* 2nd ed. Boca Raton, FL: CRC Press, 2017.

Rusnock, Paul, and Jan Šebestík. *Bernard Bolzano: His Life and Work.* Oxford, U.K.: Oxford University Press, 2019.

Spiegelhalter, David. *The Art of Statistics: How to Learn from Data.* New York: Basic Books, 2019.

Strogatz, Steven. *Infinite Powers: How Calculus Reveals the Secrets of the Universe.* Boston: Houghton Mifflin Harcourt, 2019.

Timberlake, Todd, and Paul Wallace. *Finding Our Place in the Solar System: The Scientific Story of the Copernican Revolution.* Cambridge, U.K.: Cambridge University Press, 2019.

Vanderburgh, William L. *David Hume on Miracles, Evidence, and Probability.* Lanham, MD: Rowman & Littlefield, 2019.

Williams, Talithia. *Power in Numbers: The Rebel Women of Mathematics.* New York: The Quarto Group, 2018.

Yeung, Karen, and Martin Lodge. (Eds.) *Algorithmic Regulation.* Oxford, U.K.: Oxford University Press, 2019.

The BEST WRITING on MATHEMATICS

2020

Outsmarting a Virus with Math

Steven Strogatz

In the 1980s, a mysterious disease began killing tens of thousands of people a year in the United States and hundreds of thousands worldwide. No one knew what it was, where it came from, or what was causing it, but its effects were clear—it weakened patients' immune systems so severely that they became vulnerable to rare kinds of cancer, pneumonia, and opportunistic infections. Death from the disease was slow, painful, and disfiguring. Doctors named it acquired immunodeficiency syndrome (AIDS). No cure was in sight.

Basic research demonstrated that a retrovirus was the culprit. Its mechanism was insidious: The virus attacked and infected white blood cells called helper T cells, a key component of the immune system. Once inside, the virus hijacked the cell's genetic machinery and co-opted it into making more viruses. Those new virus particles then escaped from the cell, hitched a ride in the bloodstream and other bodily fluids, and looked for more T cells to infect. The body's immune system responded to this invasion by trying to flush out the virus particles from the blood and kill as many infected T cells as it could find. In so doing, the immune system was killing an important part of itself.

The first antiretroviral drug approved to treat HIV appeared in 1987. It slowed the virus down by interfering with the hijacking process, but it was not as effective as hoped, and HIV often became resistant to it. A different class of drugs called protease inhibitors appeared in 1994. They thwarted HIV by interfering with the newly produced virus particles, keeping them from maturing and rendering them noninfectious. Though also not a cure, protease inhibitors were a godsend.

Soon after protease inhibitors became available, a team of researchers led by David Ho (a former physics major at the California Institute of Technology and so, presumably, someone comfortable with calculus) and a mathematical immunologist named Alan Perelson collaborated on a study that changed how doctors thought about HIV and revolutionized how they treated it. Before the work of Ho and Perelson, it was known that untreated HIV infection typically progressed through three stages: an acute primary stage of a few weeks, a chronic and paradoxically asymptomatic stage of up to 10 years, and a terminal stage of AIDS.

In the first stage, soon after a person becomes infected with HIV, he or she displays flulike symptoms of fever, rash, and headaches, and the number of helper T cells (also known as CD4 cells) in the bloodstream plummets. A normal T cell count is about 1,000 cells per cubic millimeter of blood; after a primary HIV infection, the T cell count drops to the low hundreds. Because T cells help the body fight infections, their depletion severely weakens the immune system. Meanwhile, the number of virus particles in the blood, known as the viral load, spikes and then drops as the immune system begins to combat the HIV infection. The flulike symptoms disappear, and the patient feels better.

At the end of this first stage, the viral load stabilizes at a level that can, puzzlingly, last for many years. Doctors refer to this level as the *set point*. A patient who is untreated may survive for a decade with no HIV-related symptoms and no lab findings other than a persistent viral load and a low and slowly declining T cell count. Eventually, however, the asymptomatic stage ends and AIDS sets in, marked by a further decrease in the T cell count and a sharp rise in the viral load. Once an untreated patient has full-blown AIDS, opportunistic infections, cancers, and other complications usually cause the patient's death within two to three years.

The key to the mystery was in the decade-long asymptomatic stage. What was going on then? Was HIV lying dormant in the body? Other viruses were known to hibernate like that. The genital herpesvirus, for example, hunkers down in nerve ganglia to evade the immune system. The chicken pox virus also does this, hiding out in nerve cells for years and sometimes awakening to cause shingles. For HIV, the reason for the latency was unknown.

In a 1995 study, Ho and Perelson gave patients a protease inhibitor, not as a treatment but as a probe. Doing so nudged a patient's body off its set point and allowed the researchers—for the first time ever—to track the dynamics of the immune system as it battled HIV. They found that after each patient took the protease inhibitor, the number of virus particles in the bloodstream dropped exponentially fast. The rate of decay was incredible: half of all the virus particles in the bloodstream were cleared by the immune system every *two days*.

Finding the Clearance Rate

Calculus enabled Perelson and Ho to model this exponential decay and extract its surprising implications. First, they represented the changing concentration of virus in the blood as an unknown function, $V(t)$, where t denotes the elapsed time since the protease inhibitor was administered. Then they hypothesized how much the concentration of virus would change, dV, in an infinitesimally short time interval, dt. Their data indicated that a constant fraction of the virus in the blood was cleared each day, so perhaps the same constancy would hold when extrapolated down to dt. Because dV/V represented the fractional change in the virus concentration, their model could be translated into symbols as the following equation:

$$dV/V = -c\,dt$$

Here the constant of proportionality, c, is the clearance rate, a measure of how fast the body flushes out the virus.

The equation above is an example of a differential equation. It relates the infinitesimal change of V (which is called the differential of V and denoted dV) to V itself and to the differential dt of the elapsed time. By applying the techniques of calculus to this equation, Perelson and Ho solved for $V(t)$ and found it satisfied:

$$\ln [V(t)/V_0] = -ct$$

Here V_0 is the initial viral load, and ln denotes a function called the natural logarithm. Inverting this function then implied:

$$V(t) = V_0 e^{-ct}$$

In this equation, e is the base of the natural logarithm, thus confirming that the viral load did indeed decay exponentially fast in the model. Finally, by fitting an exponential decay curve to their experimental data, Ho and Perelson estimated the previously unknown value of c.

For those who prefer derivatives (rates of change) to differentials (infinitesimal increments of change), the model equation can be rewritten as follows:

$$dV/dt = -cV$$

Here dV/dt is the derivative of V with respect to t. This derivative measures how fast the virus concentration grows or declines. Positive values signify growth; negative values indicate decline. Because the concentration V is positive, then $-cV$ must be negative. Thus, the derivative must also be negative, which means the virus concentration has to decline, as we know it does in the experiment. Furthermore, the proportionality between dV/dt and V means that the closer V gets to zero, the more slowly it declines.

This slowing decline of V is similar to what happens if you fill a sink with water and then allow it to drain. The less water in the sink, the more slowly it flows out because less water pressure is pushing it down. In this analogy, the volume of water in the sink is akin to the amount of virus in the body; the drainage rate is like the outflow of the virus as it is cleared by the immune system.

Having modeled the effect of the protease inhibitor, Perelson and Ho modified their equation to describe the conditions *before* the drug was given. They assumed the equation would become:

$$dV/dt = P - cV$$

In this equation, P refers to the uninhibited rate of production of new virus particles, another crucial unknown in the early 1990s. Perelson and Ho imagined that before administration of the protease inhibitor, infected cells were releasing new infectious virus particles at every moment, which then infected other cells, and so on. This potential for a raging fire is what makes HIV so devastating.

In the asymptomatic phase, however, there is evidently a balance between the production of the virus and its clearance by the immune system. At this set point, the virus is produced as fast as it is cleared. That gave new insight into why the viral load could stay the same for years. In the water-in-the-sink analogy, it is like what happens if you turn on the faucet and open the drain at the same time. The water will reach a steady-state level at which outflow equals inflow.

At the set point, the concentration of virus does not change, so its derivative has to be zero: $dV/dt = 0$. Hence, the steady-state viral load V_0 satisfies:

$$P = cV_0$$

Perelson and Ho used this simple equation to estimate a vitally important number that no one had found a way to measure before: the number of virus particles being cleared each day by the immune system. It turned out to be a *billion* virus particles a day.

That number was unexpected and truly stunning. It indicated that a titanic struggle was taking place during the seemingly calm 10 years of the asymptomatic phase in a patient's body. The immune system cleared a billion virus particles daily, and the infected cells released a billion new ones. The immune system was in a furious, all-out war with the virus and fighting it to a near standstill.

Turning Hibernation on Its Head

The following year, Ho, Perelson, and their colleagues conducted a follow-up study to get a better handle on something they could not resolve in 1995. This time they collected viral load data at shorter time intervals after the protease inhibitor was administered because they wanted to obtain more information about an initial lag they had observed in the medicine's absorption, distribution, and penetration into the target cells. After the drug was given, the team measured the patients' viral load every two hours until the sixth hour, then every six hours until day two and then once a day thereafter until day seven. On the mathematical side, Perelson refined the differential equation model to account for the lag and to track the dynamics of another important variable, the changing number of infected T cells.

When the researchers reran the experiment, fit the data to the model's predictions, and estimated its parameters again, they obtained results even more staggering than before: *10 billion* virus particles were being produced and then cleared from the bloodstream each day. Moreover, they found that infected T cells lived only about two days. The surprisingly short life span added another piece to the puzzle, given that T cell depletion is the hallmark of HIV infection and AIDS.

The discovery that HIV replication was so astonishingly rapid changed the way that doctors treated their HIV-positive patients. Previously physicians waited until HIV emerged from its supposed hibernation before they prescribed antiviral drugs. The idea was to conserve forces until the patient's immune system really needed help because the virus would often become resistant to the drugs. So it was generally thought wiser to wait until patients were far along in their illness.

Ho and Perelson turned this picture upside down. There was no hibernation. HIV and the body were locked in a pitched struggle every second of every day, and the immune system needed all the help it could get and as soon as possible after the critical early period of infection. And now it was obvious why no single medication worked for very long. The virus replicated so rapidly and mutated so quickly, it could find a way to escape almost any therapeutic drug.

Perelson's mathematics gave a quantitative estimate of how many drugs had to be used in combination to beat HIV down and keep it down. By taking into account the measured mutation rate of HIV, the size of its genome, and the newly estimated number of virus particles that were produced daily, he demonstrated mathematically that HIV was generating every possible mutation at every base in its genome many times a day. Because even a single mutation could confer drug resistance, there was little hope of success with single-drug therapy. Two drugs given at the same time would stand a better chance of working, but Perelson's calculations showed that a sizable fraction of all possible double mutations also occurred each day. Three drugs in combination, however, would be hard for the HIV virus to overcome. The math suggested that the odds were something like 10 million to one against HIV being able to undergo the necessary three simultaneous mutations to escape triple-combination therapy.

When Ho and his colleagues tested a three-drug cocktail on HIV-infected patients in clinical studies in 1996, the results were remarkable. The level of virus in the blood dropped about 100-fold in two weeks. Over the next month, it became undetectable.

This is not to say that HIV was eradicated. Studies soon afterward showed that the virus can rebound aggressively if patients take a break from therapy. The problem is that HIV can hide out. It can lie low in sanctuary sites in the body that the drugs cannot readily penetrate or lurk in latently infected cells and rest without replicating, a sneaky way of evading treatment. At any time, these dormant cells can wake up and start making new viruses, which is why it is so important for HIV-positive people to keep taking their medications, even when their viral loads are undetectable.

In 1996, Ho was named *Time* magazine's Man of the Year. In 2017, Perelson received a major prize for his "profound contributions to theoretical immunology." Both are still saving lives by applying calculus to medicine: Ho is analyzing viral dynamics, and some of Perelson's latest work helped to create treatments for hepatitis C that cure the infection in nearly every patient.

The calculus that led to triple-combination therapy did not cure HIV. But it changed a deadly virus into a chronic condition that could be managed—at least for those with access to treatment. It gave hope where almost none had existed before.

More to Explore

Rapid Turnover of Plasma Virions and CD4 Lymphocytes in HIV-1 Infection. David D. Ho et al. in *Nature*, Vol. 373, pp. 123–126, January 12, 1995.

Modelling Viral and Immune System Dynamics. Alan S. Perelson in *Nature Reviews Immunology*, Vol. 2, pp. 28–36, January 2002.

Uncertainty

PETER J. DENNING AND TED G. LEWIS

In a famous episode in the "I Love Lucy" television series—"Job Switching," better known as the chocolate factory episode—Lucy and her best-friend coworker Ethel are tasked to wrap chocolates flowing by on a conveyor belt in front of them. Each time they get better at the task, the conveyor belt speeds up. Eventually they cannot keep up, and the whole scene collapses into chaos.

The threshold between order and chaos seems thin. A small perturbation—such as a slight increase in the speed of Lucy's conveyor belt—can either do nothing or it can trigger an avalanche of disorder. The speed of events within an avalanche overwhelms us, sweeps away structures that preserve order, and robs our ability to function. Quite a number of disasters, natural or human-made, have an avalanche character—earthquakes, snow cascades, infrastructure collapse during a hurricane, or building collapse in a terror attack. Disaster-recovery planners would dearly love to predict the onset of these events so that people can safely flee and first responders can restore order with recovery resources standing in reserve.

Disruptive innovation is also a form of avalanche. Businesses hope their new products will "go viral" and sweep away competitors. Competitors want to anticipate market avalanches and side-step them. Leaders and planners would love to predict when an avalanche might occur and how extensive it might be.

In recent years, complexity theory has given us a mathematics to deal with systems where avalanches are possible. Can this theory make the needed predictions where classical statistics cannot? Sadly, complexity theory cannot do this. The theory is very good at explaining avalanches after they have happened, but generally useless for predicting when they will occur.

Complexity Theory

In 1984, a group of scientists founded the Santa Fe Institute to see if they could apply their knowledge of physics and mathematics to give a theory of chaotic behavior that would enable professionals and managers to move productively amid uncertainty. Over the years, the best mathematical minds developed a beautiful, rich theory of complex systems.

Traditional probability theory provides mathematical tools for dealing with uncertainty. It assumes that the uncertainty arises from random variables that have probability distributions over their possible values. It typically predicts the future values of the variable by computing a mean of the distribution and a confidence interval based on its standard deviation. For example, in 1962 Everett Rogers studied the adoption times of the members of a community in response to a proposed innovation (5). He found they follow a normal (bell) curve that has a mean and a standard deviation. A prediction of adoption time is the mean time bracketed by a confidence interval: for example, 68% of the adoption times are within one standard deviation of the mean, and 95% are within two standard deviations.

In 1987, researchers Per Bak, Chao Tang, and Kurt Wiesenfeld published the results of a simple experiment that demonstrated the essence of complexity theory (4). They observed a sand pile as it formed by dropping grains of sand on a flat surface. Most of the time, each new grain would settle into a stable position on the growing cone of sand. But at unpredictable moments, a grain would set off an avalanche of unpredictable size that cascaded down the side of the sand pile. The researchers measured the time intervals between avalanche starts and the sizes of avalanches. To their surprise, these two random variables did not fit any classical probability distribution, such as the normal or Poisson distributions. Instead, their distributions followed a "power law," meaning the probability of a sample of length x is proportional to x^{-k}, where k is a fixed parameter of the random process. Power law distributions have a finite mean only if $k > 2$ and variance only if $k > 3$. This means that a power law with $k \leq 2$ has no mean or variance. Its future is unpredictable. When $2 < k \leq 3$, the mean is finite but not the confidence interval. Bak et al. had discovered something different—a random process whose future could not be predicted with any confidence.

This was not an isolated finding. Most of the random processes tied to chaotic situations obey a power law with $k < 3$. For example, the appearance of new connections among web pages is chaotic. The number of web pages with x connections to other pages is proportional to $1/x^2$—the random process of accumulating links produces $1/4$ as many pages with $2x$ connections as with x connections. This was taken as both bad and good news for the Internet. The bad news is that because there are a very few "hubs"—servers hosting a very large number of connections—an attacker could shatter the network into isolated pieces by bringing down the hubs. The good news is that the vast majority of servers host few connections and thus random server failures are unlikely to shatter the network. What makes this happen is "preferential attachment"—when a new web page joins the network, it tends to connect with the most highly connected nodes already in the network. Start-up company founders try to plot strategies to bring about rapid adoption of their technologies and transform their new services into hubs.

Hundreds of processes in science and engineering follow power laws, and their key variables are unpredictable. Innovation experts believe that innovations follow a power law—the number of innovations adopted by communities of size x is proportional to x^{-2}—not good news for start-up companies hoping to predict that their innovations will take over the market.

Later Bak (1) developed a theory of unpredictability that has subsequently been copied by popular writers like Nassim Nicholas Taleb (6) and others. Bak called it *punctuated equilibrium*, a concept first proposed by Stephen Jay Gould and Niles Eldredge in 1972 (3). The idea is that new members can join a complex system by fitting into the existing structure; but occasionally, the structure passes a critical point and collapses and the process starts over. The community order that has worked for a long time can become brittle. *Avalanche* is an apt term for the moment of collapse. In the sand pile, for example, most new grains lodge firmly into a place on the pile, but occasionally one sets off an avalanche that changes the structure. On the Internet, malware can quickly travel via a hub to many nodes and cause a large-scale avalanche of disruption. In an economy, a new technology can suddenly trigger an avalanche that sweeps away an old structure of jobs and professions and establishes a new order, leaving many people stranded. Complexity theory tells us we frequently encounter systems that transition between stability and randomness.

Punctuated equilibrium appears differently in different systems because self-organization manifests in different ways. On the Internet, it may be the vulnerability to the failure of highly connected hubs. In a national highway system, it may be the collapse of maintenance as more roads are added, bringing new traffic that deteriorates old roads faster. In geology, it may be the sudden earthquake that shatters a stable fault and produces a cascade of aftershocks. In a social system, it may be the outbreak of protests when people become "fed up."

Explanations but Not Predictions

What can we learn from all this? Many systems have a strong social component, which leads to forms of preferential attachment and power laws governing the degrees of connectivity in the social network. These systems are susceptible to sudden changes of structure of unpredictable onset and extent. The best we can say is that the conditions for avalanche are present, but we cannot say with any certainty that the avalanche will actually happen or, if it does, what its extent will be. In other words, we are able to explain an avalanche after it happens, but we are profoundly unable to predict anything about it before it happens.

Earthquake preparedness is an example in nature that does not depend on humans. Seismic experts can tell us where the fault lines are and compute the probabilities of earthquake on different faults. They cannot, however, predict when an earthquake will happen or how large it will be. In effect, they are trying to predict when an earth avalanche—collapse of structure in a section of Earth's crust—will happen. Similarly, snow experts know when conditions are "ripe" for an avalanche and can call for evacuating the area. But they cannot know exactly where a snow avalanche may start, or when, or how much snow will sweep down. These experts call on people to be prepared, but few actually heed the advice and lay in supplies or make contingency plans.

Navigating in Uncertainty

Complexity researchers have turned to simulations of complex systems to see when avalanches happen and how large they are. These simulations often reveal regularities in the state spaces of those systems that can be usefully exploited to make predictions.

What are more pragmatic things we can do to cope with uncertainty? We can learn some lessons from those who must deal with disasters such as fires, earthquakes, floods, or terror attacks. Their data show that the times between events and sizes of events follow power laws and cannot be predicted. Their coping strategy boils down to preparedness, resiliency, and adaptiveness. *Preparedness* means having recovery resources standing by in case of need. *Resiliency* means rapidly bouncing back and restoring order and function. *Adaptiveness* means planning for the "new normal" that will follow recovery.

These researchers have worked out strategies to identify the situations most "ripe" for an avalanche. For instance, the power law for terror attacks shows that attacks tend to cluster in time at a given location. Thus, a future attack is more likely to occur at the same

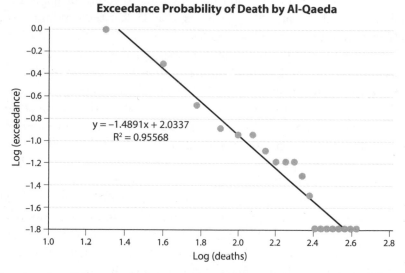

Exceedance Probability of Death by Al-Qaeda

$y = -1.4891x + 2.0337$
$R^2 = 0.95568$

Log-log plot of the exceedance versus intervals between terror attacks follows a straight line. Exceedance is the probability that an interval is greater than x (a tail of the distribution). A straight line on the log-log plot is the signature of a power law; here the slope is -1.4, telling us the tails of the distribution are a power law $y = x^{-1.4}$. Because 1.4 is less than 2, this distribution has no finite mean or standard deviation: the time to the next terror attack is unpredictable.

location than further away. The preparedness strategies include rapid mobilization of law enforcement just after an attack to counter the tendency for a new attack and identifying optimal geographic locations for positioning recovery resources and supplies. Resilience strategies include rapidly mobilizing technicians and artisans to restore broken communications and facilities. Adaptiveness strategies include scenarios and war games.

Uncertainty in Professional Work

What can we do when we find ourselves in chaotic situations and must still navigate through the uncertainty to achieve our goals?

One of the most difficult environments to navigate is the social space in which we perform our work. This space is dominated by choices that other people make beyond our control. When we propose innovations, we are likely to encounter resistance from some sectors of our community that do not want the innovation; they can be quite inventive in finding ways to block our proposals (2). When we start new projects or even companies, we do not know whether our plans are going to take off or just wither away. Even in normal everyday working environments, conflicts and contingencies suddenly arise and we must resolve them to keep moving forward.

The analogy of a surfer is useful in approaching these situations. A surfer aims to ride the waves to the shore without losing balance and being swept under. The waves can be turbulent and unpredictable. The surfer must maintain balance, ride the crests moving toward the shore, and dodge side waves and cross currents. The surfer may need to jump to a new wave when the time is right, or quickly tack to avoid an unfavorable current or wind. Thus, the surfer generates a path through the fast-changing waves.

In the social space, waves manifest as groups of people disposed to move in certain directions and not in others—sometimes the waves appear as fads or "memes," and they have a momentum that is difficult to divert. As professionals, we become aware of these waves and try to harness them to carry us toward our goals. As each surprise pops up, we instinctively look for openings into which we can move—and, more importantly, we create openings by starting conversations that assuage the concerns of those whose resistance threatens to block us.

These little deals cut a path through the potential resistance and get us to our goal.

The lesson here is that we listen for the waves, ride their momentum toward our goals, and make adjustments by creating openings in our conversations with other people. At its best, the complexity theory helps us understand when a process is susceptible to unpredictable avalanches. We move beyond the limitations of the theory by generating openings in our conversations with other people.

References

1. Bak, P. 1996. *How Nature Works: The Science of Self-Organized Criticality*. Springer-Verlag, New York.
2. Denning, P. 2018. Winning at innovation. *IEEE Computer* (Oct.), 51:10, 32–39.
3. Eldredge, N., and Gould, S. J. 1972. Punctuated Equilibria: An Alternative to Phyletic Gradualism. In *Models in Paleobiology*, T. J. M. Schopf, Ed. Freeman Cooper, San Francisco, 82–115.
4. Lewis, T. G. 2011. *Bak's Sand Pile: Strategies for a Catastrophic World*. Agile Press, Williams, CA, 382.
5. Rogers, E. M. 2003. *Diffusion of Innovations,* 5th ed. Free Press, New York.
6. Taleb, N. N. 2007, 2010. *The Black Swan: The Impact of the Highly Improbable*. Random House, New York.

The Inescapable Casino

BRUCE M. BOGHOSIAN

Wealth inequality is escalating at an alarming rate, not only within the United States, but also in countries as diverse as Russia, India, and Brazil. According to investment bank Credit Suisse, the fraction of global household wealth held by the richest 1% of the world's population increased from 42.5 to 47.2% between the financial crisis of 2008 and 2018. To put it another way, as of 2010, 388 individuals possessed as much household wealth as the lower half of the world's population combined—about 3.5 billion people; today Oxfam estimates that number as 26 individuals. Statistics from almost all nations that measure wealth in their household surveys indicate that wealth is becoming increasingly concentrated.

Although the origins of inequality are hotly debated, an approach developed by physicists and mathematicians, including my group at Tufts University, suggests that they have long been hiding in plain sight—in a well-known quirk of arithmetic. This method uses models of wealth distribution collectively known as agent-based, which begin with an individual transaction between two "agents" or actors, each trying to optimize his or her own financial outcome. In the modern world, nothing could seem more fair or natural than two people deciding to exchange goods, agreeing on a price and shaking hands. Indeed, the seeming stability of an economic system arising from this balance of supply and demand among individual actors is regarded as a pinnacle of Enlightenment thinking—to the extent that many people have come to conflate the free market with the notion of freedom itself. Our deceptively simple mathematical models, which are based on voluntary transactions, suggest, however, that it is time for a serious reexamination of this idea.

In particular, the *affine wealth model* (called thus because of its mathematical properties) can describe wealth distribution among households in diverse developed countries with exquisite precision while revealing a subtle asymmetry that tends to concentrate wealth. We believe that this purely analytical approach, which resembles an x-ray in that it is used not so much to represent the messiness of the real world as to strip it away and reveal the underlying skeleton, provides deep insight into the forces acting to increase poverty and inequality today.

Oligarchy

In 1986, social scientist John Angle first described the movement and distribution of wealth as arising from pairwise transactions among a collection of "economic agents," which could be individuals, households, companies, funds, or other entities. By the turn of the century, physicists Slava Ispolatov, Pavel L. Krapivsky, and Sidney Redner, then all working together at Boston University, as wel as Adrian Drǎgulescu, now at Constellation Energy Group, and Victor Yakovenko of the University of Maryland, had demonstrated that these agent-based models could be analyzed with the tools of statistical physics, leading to rapid advances in our understanding of their behavior. As it turns out, many such models find wealth moving inexorably from one agent to another—even if they are based on fair exchanges between equal actors. In 2002, Anirban Chakraborti, then at the Saha Institute of Nuclear Physics in Kolkata, India, introduced what came to be known as the "yard sale model," called thus because it has certain features of real one-on-one economic transactions. He also used numerical simulations to demonstrate that it inexorably concentrated wealth, resulting in oligarchy.

To understand how this happens, suppose you are in a casino and are invited to play a game. You must place some ante—say, $100—on a table, and a fair coin will be flipped. If the coin comes up heads, the house will pay you 20% of what you have on the table, resulting in $120 on the table. If the coin comes up tails, the house will take 17% of what you have on the table, resulting in $83 left on the table. You can keep your money on the table for as many flips of the coin as you would like (without ever adding to or subtracting from it). Each time you play, you will win 20% of what is on the table if the coin comes up heads, and you will lose 17% of it if the coin comes up tails. Should you agree to play this game?

You might construct two arguments, both rather persuasive, to help you decide what to do. You may think, "I have a probability of $\frac{1}{2}$ of gaining \$20 and a probability of $\frac{1}{2}$ of losing \$17. My expected gain is therefore:

$$\tfrac{1}{2} \times (+\$20) + \tfrac{1}{2} \times (-\$17) = \$1.50$$

which is positive. In other words, my odds of winning and losing are even, but my gain if I win will be greater than my loss if I lose." From this perspective, it seems advantageous to play this game.

Or, like a chess player, you might think further: "What if I stay for 10 flips of the coin? An extension of the above argument indicates that my expected winning will be $(1 + 0.015)^{10} \times \$100 = \$116.05$. This is correct, and it seems promising until I realize that I would need at least six wins to avoid a loss. Five wins and five losses will not be good enough, since the amount of money remaining on the table in that case would be

$$1.2 \times 1.2 \times 1.2 \times 1.2 \times 1.2 \times 0.83 \times 0.83 \times 0.83$$
$$\times 0.83 \times 0.83 \times \$100 = \$98.02$$

so I will have lost about \$2 of my original \$100 ante. The trouble is that, of the $2^{10} = 1,024$ possible outcomes of 10 coin flips, only 386 of them result in my winning six or more times, so the probability of that happening is only $386/1024 = 0.377$. Hence, while my reward for winning is increasing, the probability of my winning is simultaneously decreasing." As the number of coin flips increases, this problem only worsens as the following table makes clear:

Number of Coin Flips	Expected Gain	Number of Wins Required to Avoid Loss	Probability of Avoiding a Loss
10	0.1605	6	0.377
100	3.432	51	0.460
1,000	2.924×10^6	506	0.364
10,000	4.575×10^{64}	5,055	0.138
100,000	4.020×10^{646}	50,544	0.000294

While the expected gains are spectacular, the probability of realizing any profit at all plummets with the number of coin flips, until the situation becomes very much like a lottery. With a bit more work, you can uncover the reason for this: It takes about 93 wins to compensate for 91 losses, and the coin is fair, so the longer you play the more likely you are to lose. From this perspective, it seems decidedly disadvantageous to play this game.

The contradiction between the two arguments presented here may seem surprising at first, but it is well known in probability and finance. Its connection with wealth inequality is less familiar, however. To extend the casino metaphor to the movement of wealth in an (exceedingly simplified) economy, let us imagine a system of 1,000 individuals who engage in pairwise exchanges with one another. Let each begin with some initial wealth, which could be exactly equal. Choose two agents at random and have them transact, then do the same with another two, and so on. In other words, this model assumes sequential transactions between randomly chosen pairs of agents. Our plan is to conduct millions or billions of such transactions in our population of 1,000 and see how the wealth ultimately is distributed.

What should a single transaction between a pair of agents look like? People have a natural aversion to going broke, so we assume that the amount at stake, which we call Δw (Δw is pronounced "delta w"), is a mere fraction of the wealth of the poorer person, Shauna. That way, even if Shauna loses in a transaction with Eric, the richer person, the amount she loses is always less than her own total wealth. This is not an unreasonable assumption, and in fact, it captures a self-imposed limitation that most people instinctively observe in their economic lives. To begin with—just because these numbers are familiar to us—let us suppose that Δw is 20% of Shauna's wealth, w, if she wins and −17% of w if she loses. (Our actual model assumes that the win and loss percentages are equal, but the general outcome still holds. Moreover, increasing or decreasing Δw just extends the timescale so that more transactions are required before we can see the ultimate result, which remains unaltered.)

If our goal is to model a fair and stable market economy, we ought to begin by assuming that nobody has an advantage of any kind, so let us decide the direction in which Δw is moved by the flip of a fair coin. If the coin comes up heads, Shauna gets 20% of her wealth from Eric; if the coin comes up tails, she must give 17% of it to Eric. Now randomly choose another pair of agents from the total of 1,000 and do it again. In fact, go ahead and do this a million times or a billion times. What happens?

If you simulate this economy, a variant of the yard sale model, you will get a remarkable result: after a large number of transactions, one agent ends up as an "oligarch," holding practically all the wealth of the economy, and the other 999 end up with virtually nothing. It does not matter how much wealth people started with. It does not matter that all the coin flips were absolutely fair. It does not matter that the poorer agent's expected outcome was positive in each transaction, whereas that of the richer agent was negative. Any single agent in this economy could have become the oligarch—in fact, all had equal odds if they began with equal wealth. In that sense, there was equality of opportunity. But only one of them *did* become the oligarch, and all the others saw their average wealth decrease toward zero as they conducted more and more transactions. To add insult to injury, the lower someone's wealth ranking, the faster the decrease.

This outcome is especially surprising because it holds even if all the agents started off with identical wealth and were treated symmetrically. Physicists describe phenomena of this kind as "symmetry breaking" [*see* box "The Physics of Inequality"]. The very first coin-flip tranfers money from one agent to another, setting up an imbalance between the two. And once we have some variance in wealth, however minute, succeeding transactions will systematically move a "trickle" of wealth upward from poorer agents to richer ones, amplifying inequality until the system reaches a state of oligarchy.

If the economy is unequal to begin with, the poorest agent's wealth will probably decrease the fastest. Where does it go? It must go to wealthier agents because there are no poorer agents. Things are not much better for the second-poorest agent. In the long run, all par ticipants in this economy except for the very richest one will see their wealth decay exponentially. In separate papers in 2015, my colleagues and I at Tufts University and Christophe Chorro of Université Panthéon-Sorbonne provided mathematical proofs of the outcome that Chakraborti's simulations had uncovered—that the yard sale model moves wealth inexorably from one side to the other.

Does this mean that poorer agents never win or that richer agents never lose? Certainly not. Once again, the setup resembles a

The Physics of Inequality

When water boils at 100 degrees Celsius and turns into water vapor, it undergoes a phase transition—a sudden and dramatic change. For example, the volume it occupies (at a given pressure) increases discontinuously with temperature. Similarly, the strength of a ferromagnet falls to zero (*line in Figure A*) as its temperature increases to a point called the Curie temperature, T_c. At temperatures above T_c, the substance has no net magnetism. The fall to zero magnetism is continuous as the temperature approaches T_c from below, but the graph of magnetization versus temperature has a sharp kink at T_c.

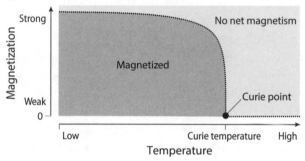

A **Phase Change in a Ferromagnet**

Conversely, when the temperature of a ferromagnet is reduced from above to below T_c, magnetization spontaneously appears where there had been none. Magnetization has an inherent spatial orientation—the direction from the south pole of the magnet to the north pole—and one might wonder how it develops. In the absence of any external magnetic field that might indicate a preferred direction, the breaking of the rotational symmetry is "spontaneous." (Rotational symmetry is the property of being identical in every orientation, which the system has at temperatures above T_c.) That is, magnetization shows up suddenly, and the direction of the magnetization is random (or, more precisely, dependent on microscopic fluctuations beyond our idealization of the ferromagnet as a continuous macroscopic system).

Economic systems can also exhibit phase transitions. When the wealthbias parameter ζ of the affine wealth model is less than the redistribution parameter χ, the wealth distribution is not even partially oligarchical (*area on the right in Figure B*). When ζ exceeds χ, however, a finite fraction of the wealth of the entire population "condenses" into the hands of an infinitesimal fraction of the wealthiest agents. The role of temperature is played by the ratio χ/ζ, and wealth condensation shows up when this quantity falls below 1. (See also color insert.)

B Phase Transition in Economic Systems

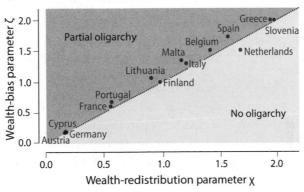

Another subtle symmetry exhibited by complex macroscopic systems is "duality," which describes a one-to-one correspondence between states of a substance above and below the critical temperature, at which the phase transition occurs. For ferromagnetism, it relates an ordered, magnetized system at temperature T below T_c to its "dual"—a disordered, unmagnetized system at the socalled inverse temperature, $(T_c)^2/T$, which is above T_c. The critical temperature is where the system's temperature and the inverse temperature cross (that is, $T = (T_c)^2/T$). Duality theory plays an increasingly important role in theoretical physics, including in quantum gravity.

Like ferromagnetism, the affine wealth model exhibits duality, as proved by Jie Li and me in 2018. A state with $\zeta < \chi$ is not a partial oligarchy, whereas a corresponding state with this relation reversed—that is, with the "temperature" χ/ζ inverted to ζ/χ—is. Interestingly, these two dual states have exactly the same wealth distribution if the oligarch is removed from the wealth-condensed economy (and the total wealth is recalculated to account for this loss).

Significantly, most countries are very close to criticality. A plot of 14 of the countries served by the European Central Bank in the $\chi - \zeta$ plane in Figure B shows that most countries lie near the diagonal. All except one (the Netherlands) lie just above the diagonal, indicating that they are just slightly oligarchical. It may be that inequality naturally increases until oligarchies begin to form, at which point political pressures set in, preventing further reduction of equality.

casino—you win some and you lose some, but the longer you stay in the casino, the more likely you are to lose. The free market is essentially a casino that you can never leave. When the trickle of wealth described earlier, flowing from poor to rich in each transaction, is multiplied by 7.7 billion people in the world conducting countless transactions every year, the trickle becomes a torrent. Inequality inevitably grows more pronounced because of the collective effects of enormous numbers of seemingly innocuous but subtly biased transactions.

The Condensation of Wealth

You might, of course, wonder how this model, even if mathematically accurate, has anything to do with reality. After all, it describes an entirely unstable economy that inevitably degenerates to complete oligarchy, and there are no complete oligarchies in the world. It is true that, by itself, the yard sale model is unable to explain empirical wealth distributions. To address this deficiency, my group has refined it in three ways to make it more realistic.

In 2017, Adrian Devitt-Lee, Merek Johnson, Jie Li, Jeremy Marcq, Hongyan Wang, and I, all at Tufts, incorporated the redistribution of wealth. In keeping with the simplicity desirable in applied mathematics models, we did this by having each agent take a step toward the mean wealth in the society after each transaction. The size of the step was some fraction χ (or "chi") of his or her distance from the mean. This is equivalent to a flat wealth tax for the wealthy (with tax rate χ per unit time) and a complementary subsidy for the poor. In effect, it transfers wealth from those above the mean to those below it. We found that this simple modification stabilized the wealth distribution so that oligarchy no longer resulted. And astonishingly, it enabled our model to match empirical data on U.S. and European wealth distribution between 1989 and 2016 to better than 2%. The single parameter χ seems to subsume a host of real-world taxes and subsidies that would be too messy to include separately in a skeletal model such as this one.

In addition, it is well documented that the wealthy enjoy systemic economic advantages, such as lower interest rates on loans and better financial advice, whereas the poor suffer systemic economic disadvantages, such as payday lenders and a lack of time to shop for the best prices. As James Baldwin once observed, "Anyone who has ever struggled with poverty knows how extremely expensive it is to be poor." Accordingly, in the same paper mentioned above, we factored in what we call wealth-attained advantage. We biased the coin flip in favor of the wealthier individual by an amount proportional to a new parameter, ζ, (or "zeta"), times the wealth difference divided by the mean wealth. This rather simple refinement, which serves as a proxy for a multitude of biases favoring the wealthy, improved agreement between the model and the upper tail (representing very wealthy people) of actual wealth distributions.

The inclusion of wealth-attained advantage also yields—and gives a precise mathematical definition to—the phenomenon of partial oligarchy. Whenever the influence of wealth-attained advantage exceeds that of redistribution (more precisely, whenever ζ exceeds χ), a vanishingly small fraction of people will possess a finite fraction, $1 - \chi/\zeta$, of societal wealth. The onset of partial oligarchy is in fact a phase transition for another model of economic transactions, as first described in 2000 by physicists Jean-Philippe Bouchaud, now at École Polytechnique, and Marc Mézard of the École Normale Supérieure. In our model, when ζ is less than χ, the system has only one stable state with no oligarchy; when ζ exceeds χ, a new, oligarchical state appears and becomes the stable state [*see* box "Winners, Losers"]. The two-parameter (χ and ζ) extended yard sale model thus obtained can match empirical data on U.S.

MEASURING INEQUALITY

In the early twentieth century, American economist Max O. Lorenz designed a useful way to quantify wealth inequality. He proposed plotting the fraction of wealth held by individuals with wealth less than w against the fraction of individuals with wealth less than w. Because both quantities are fractions ranging from 0 to 1, the plot fits neatly into the unit square. Twice the area between Lorenz's curve and the diagonal is called the Gini coefficient, a commonly used measure of inequality.

Let us first consider the egalitarian case. If every individual has exactly the same wealth, any given fraction of the population has precisely that fraction of the total wealth. Hence, the Lorenz curve is the diagonal (*green line in Figure A*), and the Gini coefficient is 0. In contrast, if one oligarch has all the wealth and everybody else has nothing, the poorest fraction f of the population has no wealth at all for any value of f that is less than 1, so the Lorenz curve is pegged to 0. But when f equals 1, the oligarch is included, and the curve suddenly jumps up to 1. The area between this Lorenz curve (*orange line on the right of the x-y axis*) and the diagonal is half the area of the square, or ½, and hence the Gini coefficient is 1. (See also color insert.)

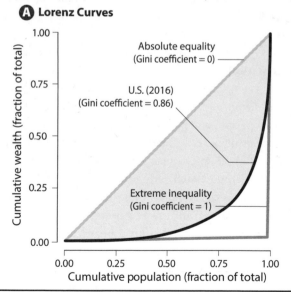

Ⓐ Lorenz Curves

Absolute equality
(Gini coefficient = 0)

U.S. (2016)
(Gini coefficient = 0.86)

Extreme inequality
(Gini coefficient = 1)

Cumulative wealth (fraction of total)

Cumulative population (fraction of total)

In sum, the Gini coefficient can vary from 0 (absolute equality) to 1 (oligarchy). Unsurprisingly, reality lies between these two extremes. The red curved line shows the actual Lorenz curve for U.S. wealth in 2016, based on data from the Federal Reserve Bank's Survey of Consumer Finances. Twice the shaded area (*yellow*) between this curve and the diagonal is approximately 0.86—among the highest Gini coefficients in the developed world.

Figure B shows the fit between the affine wealth model (AWM) and actual Lorenz curves for the United States in 1989 and 2016 and for Germany and Greece in 2010. The data are from the Federal Reserve Bank (U.S., as mentioned above) and the European Central Bank (Germany and Greece). The discrepancy between the AWM and Lorenz curves is less than a fifth of a percent for the United States and less than a third of a percent for the European countries. The Gini coefficient for the United States (*shown in plot*) increased between 1989 and 2016, indicating a rise in inequality.

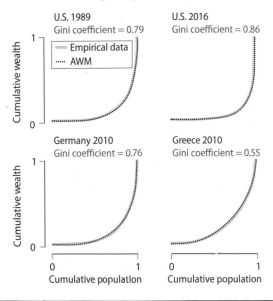

B Empirical Data compared to the Affine Wealth Model (AWM)

U.S. 1989
Gini coefficient = 0.79

— Empirical data
···· AWM

U.S. 2016
Gini coefficient = 0.86

Germany 2010
Gini coefficient = 0.76

Greece 2010
Gini coefficient = 0.55

Cumulative wealth

Cumulative population

and European wealth distribution between 1989 and 2016 to within 1 to 2%.

Such a phase transition may have played a crucial role in the condensation of wealth following the breakup of the Soviet Union in 1991. The imposition of what was called shock therapy economics on the former states of the USSR resulted in a dramatic decrease of wealth redistribution (that is, decreasing χ) by their governments and a concomitant jump in wealth-attained advantage (increasing ζ) from the combined effects of sudden privatization and deregulation. The resulting decrease of the "temperature" χ/ζ threw the countries into a wealth-condensed state, so that formerly communist countries became partial oligarchies almost overnight. To the present day, at least 10 of the 15 former Soviet republics can be accurately described as oligarchies.

As a third refinement, in 2019 we included negative wealth—one of the more disturbing aspects of modern economies—in our model. In 2016, for example, approximately 10.5% of the U.S. population was in net debt because of mortgages, student loans, and other factors. So we introduced a third parameter, κ (or "kappa"), which shifts the wealth distribution downward, thereby accounting for negative wealth. We supposed that the least wealth the poorest agent could have at any time was $-S$, where S equals κ times the mean wealth. Prior to each transaction, we loaned wealth S to both agents so that each had positive wealth. They then transacted according to the extended yard sale model, described earlier, after which they both repaid their debt of S.

The three-parameter (χ, ζ, κ) model thus obtained, called the affine wealth model, can match empirical data on U.S. wealth distribution to less than a sixth of a percent over a span of three decades. (In mathematics, the word "affine" describes something that scales multiplicatively and translates additively. In this case, some features of the model, such as the value of Δw, scale multiplicatively with the wealth of the agent, whereas other features, such as the addition or subtraction of S, are additive translations or displacements in "wealth space.") Agreement with European wealth-distribution data for 2010 is typically better than a third to a half of a percent [*see* box "The Physics of Inequality"].

To obtain these comparisons with actual data, we had to solve the "inverse problem." That is, given the empirical wealth distribution, we had to find the values of χ, ζ, and κ at which the results of our model most closely matched it. As just one example, the 2016 U.S. household wealth distribution is best described as having $\chi = 0.036$, $\zeta = 0.050$, and $\kappa = 0.058$. The affine wealth model has been applied to empirical data from many countries and epochs. To the best of our knowledge, it describes wealth-distribution data more accurately than any other existing model.

Trickle Up

We find it noteworthy that the best-fitting model for empirical wealth distribution discovered so far is one that would be completely unstable without redistribution rather than one based on a supposed equilibrium of market forces. In fact, these mathematical models demonstrate that far from wealth trickling down to the poor, the natural inclination of wealth is to flow upward, so that the "natural" wealth distribution in a free-market economy is one of complete oligarchy. It is only redistribution that sets limits on inequality.

The mathematical models also call attention to the enormous extent to which wealth distribution is caused by symmetry breaking, chance, and early advantage (from, for example, inheritance). And the presence of symmetry breaking puts paid to arguments for the justness of wealth inequality that appeal to "voluntariness"—the notion that individuals bear all responsibility for their economic outcomes simply because they enter into transactions voluntarily—or to the idea that wealth accumulation must be the result of cleverness and industriousness. It is true that an individual's location on the wealth spectrum correlates to some extent with such attributes, but the overall shape of that spectrum can be explained to better than 0.33% by a statistical model that completely ignores them. Luck plays a much more important role than it is usually accorded, so that the virtue commonly attributed to wealth in modern society—and, likewise, the stigma attributed to poverty—is completely unjustified.

Moreover, only a carefully designed mechanism for redistribution can compensate for the natural tendency of wealth to flow from the poor to the rich in a market economy. Redistribution is often confused with taxes, but the two concepts ought to be kept quite separate. Taxes flow from people to their governments to finance those governments' activities. Redistribution, in contrast, may be implemented by governments, but it is best thought of as a flow of wealth from people to people to compensate for the unfairness inherent in market economics. In a flat redistribution scheme, all those possessing wealth below the mean would receive net funds, whereas those above the mean would pay. And precisely because current levels of inequality are so extreme, far more people would receive than would pay.

Given how complicated real economies are, we find it gratifying that a simple analytical approach developed by physicists and mathematicians describes the actual wealth distributions of multiple nations with unprecedented precision and accuracy. Also rather curious is that these distributions display subtle but key features of complex physical systems. Most important, however, the fact that a sketch of the free market as simple and plausible as the affine wealth model gives rise to economies that are anything but free and fair should be both a cause for alarm and a call for action.

More to Explore

A Nonstandard Description of Wealth Concentration in Large-Scale Economies. Adrian Devitt-Lee et al. in *SIAM Journal on Applied Mathematics*, (78, 2), 996–1008, March 2018.
The Affine Wealth Model: An Agent-Based Model of Asset Exchange That Allows for Negative-Wealth Agents and Its Empirical Validation. Jie Li et al. in *Physica A: Statistical Mechanics and Its Applications*, 516, 423–442, February 2019.

Resolving the Fuel Economy Singularity

STAN WAGON

Consider this classic puzzle: Charlie runs four miles at A miles per hour (mph) and then four miles at B mph. What is his average speed for the eight miles?

At a glance, this puzzle seems the same as this problem: Diane runs at a rate of C minutes per mile for the first four miles and D minutes per mile for the second four. What is her average rate?

But this second average is easy to compute using the familiar *arithmetic mean*: $(C + D)/2$. You can check that Charlie's average speed is the *harmonic mean* of A and B,

$$\frac{2}{\frac{1}{A} + \frac{1}{B}}$$

This idea applies to other rates as well. Suppose a family has two vehicles, one getting 20 miles per gallon (MPG) and the other 40 MPG, and they each cover the same distance in a year. Then the combined fuel economy is the harmonic mean of 20 and 40, 26.7 MPG.

I recently purchased a Chevrolet Bolt, an electric vehicle (EV) with a 60 kilowatt-hour (kWh) battery and a 238-mile range, and realized that this unexpected behavior of rates has surprising ramifications in measuring energy economy—not only for gasoline-powered cars, but also especially for EVs. The new feature of EVs (also hybrids) is that energy consumption can be both positive and negative.

The MPG Illusion

American drivers are familiar with the MPG measure of fuel efficiency; 40 MPG is a typical value. Other countries use liters per 100 kilometers

(40 MPG is 5.9 L/100 km). The choice of units is not important, but the question of which unit to put in the denominator is.

Using MPG to measure fuel economy can skew our thinking in terms of improved efficiency. Consider this scenario: Alice buys a new SUV, improving her usage rate from 15 to 20 MPG, while Bob trades in his car, improving his usage rate from 40 to 50 MPG. Who will save more fuel? Alice will save 1.67 gallons for 100 miles ($\frac{100}{15} - \frac{100}{20} = \frac{5}{3}$), whereas Bob saves only a half-gallon over the same distance (Figure 1). In order to match Alice's gain, Bob would need a vehicle that gets 120 MPG. This phenomenon, caused by ignoring the complexity of reciprocals, is known as the *MPG illusion* (Larrick and Soll [1]).

The difference between Alice's and Bob's upgrades is much easier to see when the distance is in the denominator: Using the gallons-per-100-miles (which we abbreviate to GPM) measure, Alice improves from 6.67 to 5 GPM, while Bob improves from 2.5 to 2 GPM.

For a more extreme example, if Alice has a truck, which she improves from 10 to 14 MPG, and Bob has a 40-MPG car, it is impossible for him to upgrade his car to match her improvement. Even a perpetual motion vehicle using no fuel would not match the truck's fuel savings!

Larrick and Soll write, "Relying on linear reasoning about MPG leads people to undervalue small improvements on inefficient vehicles. We believe this general misunderstanding of MPG has implications for both public policy and research on environmental decision making."

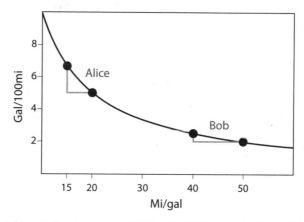

FIGURE 1. The relation between MPG and GPM ($y = 100/x$). Bob's Δx is twice Alice's, but his Δy (the fuel savings) is less than one-third of Alice's.

Because of these issues, the label placed by the Environmental Protection Agency on new cars now shows GPM and MPG.

MPG is indeed useful for measuring range and is in no danger of being abandoned. But the MPG illusion is only one of several serious problems with putting fuel in the denominator.

The MPG Paradox

The potential to negatively affect public policy decisions is not merely hypothetical. The government imposes penalties for automakers whose fleets do not satisfy the Corporate Average Fuel Economy (CAFE) standards. Steven Tenn and John Yun observe that the penalty for a carmaker might *increase* when it adds a fuel-efficient car to the fleet [2]. This surely is not the intent of the law; it appears that the people who devised the penalty formula were blinded by the traditional reliance on MPG and did not understand how fuel economy really works.

Suppose the government target is T MPG, and a manufacturer makes n cars of several models that have an average fuel economy of H MPG, which was computed (correctly) using the harmonic mean. Then the CAFE penalty is $Pn(T - H)$, where P is the penalty per car, in dollars.

Here is an example of the paradox that arises from the use of the MPG-based penalty. Suppose a company makes 1,000 cars, each of which gets 30 MPG. Suppose the government target is 50 MPG and the penalty coefficient is \$50. Then the penalty is \$50(1,000)(50 - 30) = \$1,000,000. Now suppose the fleet is enhanced by the addition of a single new car attaining 70 MPG. This improves the average fuel economy to

$$\frac{1,001}{\frac{1,000}{30} + \frac{1}{70}} = 30.017 \text{ MPG}$$

but it does not offset the fact that a penalty must be paid on all cars. The penalty is now \$1,000,142.

Quoting Tenn and Yun, "This result is driven by the fact that CAFE's regulatory tax punishes fuel efficient, as well as inefficient, vehicles. With such a blunt regulatory instrument, it is not surprising that CAFE can create peculiar incentives."

A better alternative is the penalty $P_0 \Sigma(G_i - T_0)$, where T_0 is the GPM target and G_i is the GPM rating of the ith car [3]. This approach treats each car as a separate entity. The preceding example becomes $T_0 = 2$, $P_0 = \$750$, $n = 1,000$, and G_i is 100/30 for each noncompliant car and 100/70 for the new car. The efficient car reduces the total penalty by $\$750 \ (2 - \frac{100}{70}) = \$429$.

The MPG Singularity

When coasting in a gasoline-powered car, fuel consumption can be zero; division by that number means that the instantaneous MPG is infinite. The solution of using a large number to replace ∞ is often used, but that causes additional problems in EVs that wish to present charts, not just single numbers. In an EV, fuel is measured in kWh and consumption in miles per kWh (similar to MPG) or kWh per 100 miles (similar to GPM). The Bolt presents the driver with a screen showing energy consumption over the past 50 miles, and the Tesla S does the same over 30 miles.

A driver needs to know the average fuel consumption during a trip, as it is critical that he or she get home (or to a charging station) before the battery dies. In my case, with a 60-kWh battery, if I start out on a 230-mile drive I want to be sure that my average consumption is very close to 4 miles per kWh (mi/kWh).

Let $k(x)$ denote the kWh used in the first x miles of a trip. Figure 2 shows a graph of $k(x)$ in which there is a long downhill between miles

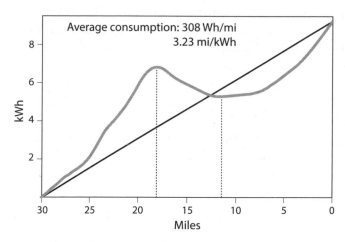

FIGURE 2. Energy used on a trip with a long downhill starting 12 miles into the trip.

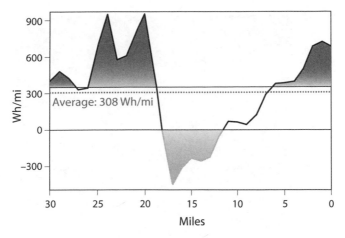

FIGURE 3. Tesla's consumption chart. The dashed line is the average; the solid line is the EPA rating of 350 Wh/mi. The green region (below the 0 line) indicates regeneration of energy into the battery. See also color insert.

12 and 18.5; on that stretch, energy regeneration decreases the total energy consumed.

The Tesla and Bolt do not show the graph in Figure 2. Tesla's display resembles Figure 3; it shows the consumption rate—the derivative $k'(x)$. It also shows the average rate, $k(30)/30$ (multiplied by 1,000 for Wh instead of kWh), as a number and as a dashed line. The units, Wh/mi, are analogous to the GPM measure.

The Bolt uses the MPG-like form mi/kWh and therefore uses the reciprocal, $1/k'(x)$. But the reciprocal creates a big problem. Because k' can be 0—this happens twice in the data of Figures 2 and 3—we can have singularities, as shown in Figure 4.

The red line (near the 0 line) in Figure 4 shows the overall average. Note that the average rate is a harmonic average of the graph. It is simply $30/k(30) = 3.25$ mi/kWh, but if f is the function $1/k'$, in Figure 4, it is

$$\frac{1}{\frac{1}{30}\int \frac{1}{f}}$$

the continuous version of the harmonic mean. The precise location of the red line is impossible to discern in Figure 4.

For a reason explained in the sidebar, Chevrolet wishes never to show negative numbers. They use the expedient of replacing any negative

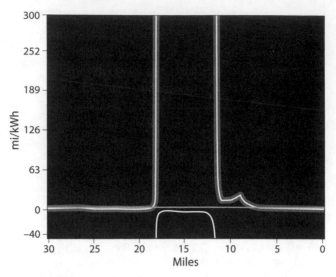

FIGURE 4. The graph of $1/k'$ is white, the average value (3.25 mi/kWh) is red, and the blue curve uses 252 as a proxy for any value that is negative or larger than 252. See also color insert.

MPG, MPGe, AND MPG$_{ghg}$

The Environmental Protection Agency introduced the miles per gallon equivalent (MPGe) so that one can evaluate an EV using the familiar MPG metric.

A gallon of gasoline has an energy content of 33.7 kWh. So if a car gets 3.97 mi/kWh (the Bolt's rating), that would translate to $(3.97)(33.7) \approx 134$ MPGe. But battery charging is not perfectly efficient: It takes about 1.12 kWh of charging electricity to put a kWh into the battery. Hence, 3.97 miles per battery-kWh translates to 3.54 miles per purchased-kWh. So the MPGe is really $(3.54)(33.7) \approx 119$ miles per virtual gallon. This explains why the Bolt's range is given as 238 miles from the 60-kWh battery, but its MPGe, at 119, is much lower than the 238 value naively indicates.

It is important to compare vehicles in a more holistic way in terms of total production of greenhouse gases, both in manufacturing and driving. For an EV, a key issue is how the electricity is generated, and this varies tremendously by region. A comprehensive study introduced MPG_{ghg} to make this comparison (Rachael Nealer, David Reichmuth, and Don Anair's 2015 report "Cleaner Cars from Cradle to Grave," www.ucsusa.org/EVlifecycle).

For example, if electricity comes from burning natural gas, then an EV might rate 58 MPG_{ghg}, meaning that a traditional car getting 58 MPG would be equivalent to that electric car in terms of total greenhouse gas production. Such a rating depends on many things, such as how exactly the batteries and the car engine were manufactured, but the dominant factor is the source of the electricity.

Multiplication Is Not as Associative as We Think

We know that $\frac{-\text{mile}}{\text{kWh}}$ and $\frac{\text{kWh}}{-\text{mile}}$ are mathematically identical, but the perceived meanings are very different. A rate of five negative miles per kWh will be generally incomprehensible, eliciting responses such as, "What is a negative mile?" and "I'll never drive five miles in reverse." But "five miles gives one negative kWh" makes perfect sense: One must drive five miles to generate a stored kWh of energy.

Using reciprocal units, this rate is -0.2 kWh/mi, and the meaning is clear: a fifth of a kWh of regeneration for each mile driven.

So while we take multiplicative associativity for granted, it is not really as universal a law as we think when perception is taken into account.

number or number larger than 252 with 252. That succeeds in eliminating the troublesome singularities! And in fact, the resulting graph (the blue line in Figure 4) is not an unreasonable approximation to the true data.

The Bolt docs not show the graph in Figure 4; instead, it displays a bar chart based on five-mile intervals, as in Figure 5. Each bar height in the bar chart can be computed exactly as, for example, $5/(k(20) - k(15))$. But the move to the bar chart does not address the problem of the

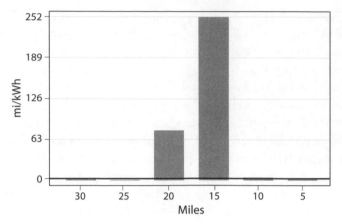

FIGURE 5. The Bolt's bar chart reduces the data of Figure 4 to five-mile intervals. The red line near 0 is the average. See also color insert.

impossible-to-interpret red line. This is inevitable, given the choice of units and the use of 252, because the chart will very often have a large vertical range.

For a bar chart based on k', the average is the arithmetic mean of the bar heights—or equivalently, their total area. But because the Bolt's chart uses $1/k'$, the average is approximated by the harmonic mean of the heights. So it has little connection to the area of the bars.

In fact, the Bolt's designers do not place the average line at the correct value (red line near 0 in Figure 5). Instead, the red line is shown at the average of the bar heights. Therefore, in the actual Bolt screen using the data of Figure 5, the red line would be at 56.6 mi/kWh as opposed to the correct 3.23; this is a 1,652% error. Chevrolet did this deliberately on the assumption that the driver would want to see the average of the green bars, but the consequence is that the driver must deal with a very incorrect average.

I was able to arrange a conversation with some Bolt engineers and emphasized that one should not present wildly inaccurate data. I believe they will rethink this approach in future models. This example shows the difficulty of using a classic MPG-like scale in EVs.

Eliminating the Singularity

A solution to the singularity problem that works with both MPG and mi/kWh is to use a nonlinear labeling. For an EV, one can use a

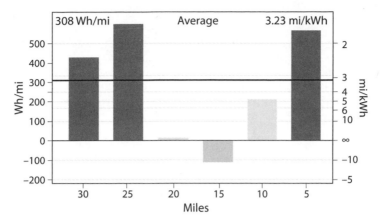

FIGURE 6. The consumption rate shown by a method that avoids the singularity and has a reasonable scale in all situations. See also color insert.

kWh/100 mi scale for the bars, but with both units for the labels as in Figure 6. Then one can read the average using either set of units (Wh/mi are used to avoid fractions).

An advantage of this approach is that the scale need never change. Going from −200 to 600 Wh/mi covers almost all situations. The singularity makes its presence felt in the ∞ that labels the horizontal axis; but it is of no consequence. In this scheme, lower means less energy consumption throughout the range, and one can represent mi/kWh values of essentially any size without sacrificing legibility for the values of interest, typically between 2 and 6 mi/kWh. Another advantage of this method is that the average is located at the average of the bar heights and so is identical to what the driver senses it should be.

Both mi/kWh (likewise MPG) and kWh/100 mi (GPM) have their places. Drivers in the United States generally want to see the usage rate in mi/kWh. The chart of Figure 6 allows this method by making use of the simple scale of the kWh/100 mi method. But the illusion, paradox, and singularity that arise from the popularity of the MPG system indicate that distance belongs in the denominator and that the GPM and kWh/100 mi measures really are superior.

Acknowledgments

The author is grateful to Paul Campbell and Jim Tilley for many helpful comments.

Notes

1. Richard Larrick and Jack Soll. "The MPG Illusion." *Science* 320 [June 20, 2008]: 1593–1594 or www.mpgillusion.com/p/what-is-mpg-illusion.html.

2. Steven Tenn and John Yun. "When Adding a Fuel Efficient Car Increases an Automaker's CAFE Penalty." *Managerial and Decision Economics* 26, 1 [2005]: 51–54.

3. Carolyn Fischer. "Let's Turn CAFE Regulation on Its Head." *Resources for the Future* [2009].

The Median Voter Theorem:
Why Politicians Move to the Center

Jørgen Veisdal

In public choice economics, the median voter theorem states that

> A majority rule voting system will select the outcome most preferred by the median voter.[1]

In other words, the voter in the middle of the probability distribution picks the winner of the election. The prediction of the model, therefore, is what intuitively seems questionable (given today's politics), namely that:

> Candidates position themselves around the center.

The theorem rests on two core assumptions:

- Candidates and/or parties may be placed along a one-dimensional political spectrum; and
- Voters' preferences are single-peaked, meaning that voters have one alternative they prefer over the other.

History

The dynamics and predictions of the median voter theorem made their first appearance in economist Harold Hotelling's[2] legendary 1929 paper *Stability in Competition*,[3] in which Hotelling in passing notes that political candidates' platforms seem to converge during majoritarian elections. His paper regards the positioning of stores by two sellers along a line segment, in which buyers are uniformly distributed. The prediction of his model, now known simply as "Hotelling's law"[4] is that in many markets it is rational for producers to make their products as similar as possible, the so-called *"principle of minimum differentiation."*

The formal analysis of the principle of Hotelling's law in majority voting systems was provided in a related 1948 paper entitled *On the Rationale of Group Decision-making* by economist Duncan Black.[5] Anthony Downs,[6] inspired by Adam Smith, further expanded on Black's work in his 1957 book *An Economic Theory of Political Action in a Democracy.*

A Simple Voting Model

The principle of the median voter theorem is successfully illustrated in a game theory lecture by Yale economist Ben Polak,[7] which is available through the Open Yale Courses website.[8] His very simple model goes as follows:

A SIMPLE VOTING MODEL (POLAK 2007)

Two candidates choose their positions along a political spectrum from 1 to 10:

—   —   —   —   —   —   —   —   —   —

1   2   3   4   5   6   7   8   9   10

Imagine perhaps positions 1–3 being left-wing positions and positions 8–10 being right-wing positions. Assume that there are 10% of voters in each position (a uniform distribution of voters). Assume also that voters choose the closest candidate to their position. If there is a tie, votes are split 50%/50% between the two candidates.

The question for both candidates is where they should position themselves along the political spectrum 1–10 in order to maximize their shares of the vote.

First, as candidate A, let us assume that candidate B places himself or herself at the extreme left, in position 1. The question we need to ask ourselves is then whether position 2 is a better strategy for maximizing our share of the vote. Formally, we're asking if $u_A(2,1) > u_A(1,1)$, i.e., whether the utility we obtain from choosing position 2 is greater than that we obtain if we choose position 1, given our assumption that the other candidate chooses position 1. We know that each position potentially holds 10% of the votes and if both choose the same position, we split the election. We—in other words—know that if both choose position 1, (1,1), both get 50% of the votes. If we as candidate A choose position 2, however, (2,1), candidate B gets the 10% of votes in position 1, but we get the remaining 90% of votes in positions 2–10 as the result of our assumption that voters choose their closest candidate:

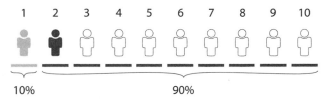

$u(1,2)$: Voter shares if candidate A chooses position 1 when candidate B chooses position 2. See also color insert.

We can conclude that if candidate B chooses position 1, choosing position 2 is better for us as candidate A than choosing position 1 (90% vs. 50% of votes).

Next, again as candidate A, let us assume that candidate B places herself or himself at the second most extreme left, in position 2. Is choosing position 2 still strictly better for us than choosing position 1? Formally, we're asking whether $u_A(2,2) > u_A(1,2)$. Immediately, since both are choosing the same position (2,2), we know that our payoff for choosing position 2 is 50%, i.e., splitting the election. If we choose position 1 against position 2, we find ourselves in the inverse of the latter scenario (in the figure above) and can expect a payoff of 10%. We must therefore conclude that if candidate B chooses position 2, as candidate A we are still better off choosing position 2 (50% > 10% of votes), than choosing position 1.

Next, again as candidate A, let us assume that candidate B places himself or herself at the third most extreme left, in position 3. Is choosing position 2 still strictly better for us than choosing position 1? Formally, now we're asking whether $u_A(2,3) > u_A(1,3)$. In the first of these scenarios, we get all the voters in position 1 plus all the voters in position 2 (a total of 20%), while candidate B gets all the voters in positions 3–10 (80%). In the second scenario, we get all the voters in position 1 plus half the voters in position 2 (since we are equally far from position 2 as candidate B is, a total of 15%). Either way, we see that position 2 is still strictly better for us as candidate A than position 1 (20% > 15% of votes):

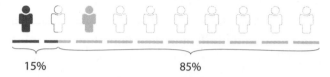

$u(1,3)$: Voter shares if candidate A chooses position 1 when candidate B chooses position 3. See also color insert.

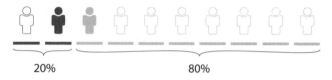

$u(2,3)$: Voter shares if candidate A chooses position 2 when candidate B chooses position 3. See also color insert.

Continuing for all strategies against choosing position 1, we will find that position 2 always yields a higher payoff. Formally, we say that the strategy of choosing position 2 *strictly dominates* the strategy of choosing position 1, for both candidates, no matter what the other candidate chooses. We can hence eliminate position 1 as a viable strategy, since it is never better than position 2 for either candidate. By considerations of symmetry, we can do the same for position 10 vs. position 9:

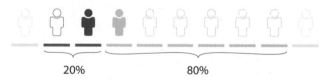

$u(2,3)$ in the reduced political spectrum. See also color insert.

Having eliminated positions 1 and 10 as viable strategies for maximizing our number of votes, we can next perform a similar analysis, but now with a reduced model consisting of positions 2–9. Unexpectedly, an analysis of position 2 vs. 3 (and 9 vs. 8) will yield the same result, and so we may delete strategies 2 and 9 as well, and so on. In the end, our model will have reduced to two viable positions, 5 and 6; let's call them "left and right."

50% 50%

$u(5,6) = u(6,5)$ in reduced political spectrum after eliminating dominated strategies. See also color insert.

An analysis of this reduced model will find that if we expect candidate B to position herself or himself on the right, it is rational for us to position ourselves on the left. If we expect candidate B to position himself or herself on the left, it is rational for us to position ourselves on the right:

50%   50%

$u(5,6) = u(6,5)$: Voter shares if candidate A chooses position 5 and candidate B chooses position 6. See also color insert.

Our finding, in other words, is the prediction of the median voter theorem, namely that:

Candidates position themselves around the center.

Clean as it looks, as with any model, our model depicting the essence of the median voter theorem is ripe with limitations, including (but not limited to) the following:

- Voters are in reality *not* evenly distributed (10% at each position);
- There are often more than two candidates in an election;
- Real candidates cannot simply "position" themselves—their positioning has to be believable; and
- Even in referendums, there is often more than one issue being voted on.

Notes

1. Holcombe, Randall G. (2006). *Public Sector Economics: The Role of Government in the American Economy.* Pearson, London, p 155. https://en.wikipedia.org/wiki/Median_voter_theorem.

2. https://en.wikipedia.org/wiki/Harold_Hotelling#Spatial_economics.

3. Hotelling, Harold. (1929). "Stability in Competition." *Economic Journal* 39, 153 (1929): 41–57.

4. https://en.wikipedia.org/wiki/Hotelling%27s_law.

5. https://en.wikipedia.org/wiki/Duncan_Black.

6. https://en.wikipedia.org/wiki/Anthony_Downs.

7. https://en.wikipedia.org/wiki/Ben_Polak.

8. https://oyc.yale.edu/economics/econ-159.

The Math That Takes Newton
into the Quantum World

JOHN BAEZ

In my 50s, too old to become a real expert, I have finally fallen in love with algebraic geometry. As the name suggests, this is the study of geometry using algebra. Around 1637, René Descartes laid the groundwork for this subject by taking a plane, mentally drawing a grid on it, as we now do with graph paper, and calling the coordinates x and y. We can write down an equation like $x^2 + y^2 = 1$, and there will be a curve consisting of points whose coordinates obey this equation. In this example, we get a circle!

It was a revolutionary idea at the time because it let us systematically convert questions about geometry into questions about equations, which we can solve if we are good enough at algebra. Some mathematicians spend their whole lives on this majestic subject. But I never really liked it much until recently—now that I have connected it to my interest in quantum physics.

As a kid, I liked physics better than math. My uncle Albert Baez, father of the famous folk singer Joan Baez, worked for UNESCO, helping developing countries with physics education. My parents lived in Washington, D.C. Whenever my uncle came to town, he would open his suitcase, pull out things like magnets or holograms, and use them to explain physics to me. This was fascinating. When I was 8, he gave me a copy of the college physics textbook he wrote. While I could not understand it, I knew right away that I *wanted* to. I decided to become a physicist, and my parents were a bit worried because they knew physicists needed mathematics, and I did not seem very good at that. I found long division insufferably boring and refused to do my math homework, with its endless repetitive drills. But later, when I realized that by fiddling around with equations I could learn about the universe, I was hooked. The mysterious symbols seemed like magic spells. And in a way, they are. Science is the magic that actually works.

In college I majored in math and became curious about theoretical physicist Eugene Wigner's question about the "unreasonable effectiveness" of mathematics: Why should our universe be so readily governed by mathematical laws? As he put it, "The miracle of the appropriateness of the language of mathematics for the formulation of the laws of physics is a wonderful gift which we neither understand nor deserve." As a youthful optimist, I felt these laws would give us a clue to the deeper puzzle: why the universe is governed by mathematical laws in the first place. I already knew that there was too much math to learn it all, so in grad school, I tried to focus on what mattered to me. And one thing that did *not* matter to me was algebraic geometry.

How could any mathematician *not* fall in love with algebraic geometry? Here is why: In its classic form, this subject considers only *polynomial* equations—equations that describe not just curves, but also higher dimensional shapes called "varieties." So, $x^2 + y^2 = 1$ is fine, and so is $x^{43} - 2xy^2 = y^7$, but an equation with sines or cosines, or other functions, is out of bounds—unless we can figure out how to convert it into an equation with just polynomials. As a graduate student, this seemed like a terrible limitation. After all, physics problems involve plenty of functions that are not polynomials.

Why does algebraic geometry restrict itself to polynomials? Mathematicians study all sorts of functions, but while they are very important, at some level their complications are only a distraction from the fundamental mysteries of the relation between geometry and algebra. By restricting the breadth of their investigations, algebraic geometers can dig deeper into these mysteries. They have been doing this for centuries, and by now their mastery of polynomials is truly staggering: Algebraic geometry has grown into a powerful tool in number theory, cryptography, and many other subjects. But for its true devotees, it is an end in itself.

I once met a grad student at Harvard, and I asked him what he was studying. He said one word, in a portentous tone: "Hartshorne." He meant Robin Hartshorne's textbook *Algebraic Geometry*, published in 1977. Supposedly an introduction to the subject, it is actually a very hard-hitting tome. Consider Wikipedia's description:

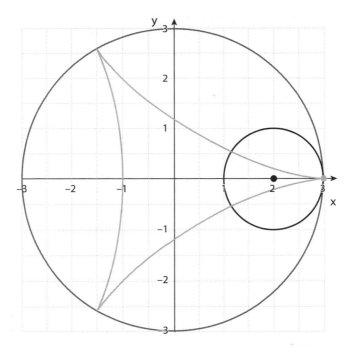

There's a polynomial for that: We can describe many interesting curves with just polynomials. For example, roll a circle inside a circle three times as big. You get a curve with three sharp corners called a "deltoid." It's not obvious that you can describe this using a polynomial equation, but you can. The great mathematician Leonhard Euler dreamed this up in 1745. Image by Sam Derbyshire.

The first chapter, titled "Varieties," deals with the classical algebraic geometry of varieties over algebraically closed fields. This chapter uses many classical results in commutative algebra, including Hilbert's Nullstellensatz, with the books by Atiyah–Macdonald, Matsumura, and Zariski–Samuel as usual references.

If you cannot make heads or tails of this . . . well, that is exactly my point. To penetrate even the first chapter of Hartshorne, you need quite a bit of background. To read Hartshorne is to try to catch up with centuries of geniuses running as fast as they could.

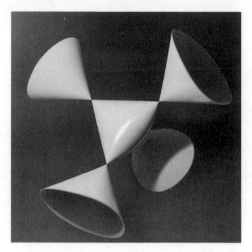

Fame, cubed: This is Cayley's nodal cubic surface. It is famous because it is the variety with the most nodes (those pointy things) that is described by a cubic equation. The equation is $(xy + yz + zx)(1 - x - y - z) + xyz = 0$, and it's called "cubic" because we're multiplying at most three variables at once. Image by Abdelaziz Nait Merzouk. See also color insert.

One of these geniuses was Hartshorne's thesis advisor, Alexander Grothendieck. From about 1960 to 1970, Grothendieck revolutionized algebraic geometry as part of an epic quest to prove the Weil Conjectures, which relate varieties to solutions of certain problems in number theory. Grothendieck guessed that the Weil Conjectures could be settled by strengthening and deepening the link between geometry and algebra. He had a concrete idea for how this should turn out. But making this idea precise required a huge amount of work. To carry it out, he started a seminar. He gave talks almost every day and enlisted the help of some of the best mathematicians in Paris.

Working nonstop for a decade, they produced tens of thousands of pages of new mathematics, packed with mind-blowing concepts. In the end, using these ideas, Grothendieck succeeded in proving all the Weil Conjectures except the final, most challenging one. A student of his polished that one off, much to Grothendieck's surprise.

During his most productive years, even though he dominated the French school of algebraic geometry, many mathematicians considered Grothendieck's ideas "too abstract." This sounds a bit strange, given

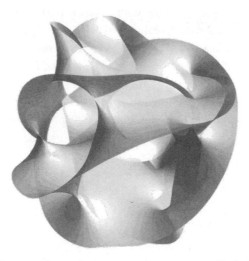

I'm all curled up: A slice of one particular variety, called a "quintic three-fold," that can be used to describe the extra curled-up dimensions of space in string theory. See also color insert.

how abstract *all* mathematics is. What is inarguably true is that it takes time and work to absorb his ideas. As a grad student, I steered clear of them since I was busy struggling to learn physics: There, too, centuries of geniuses have been working full-speed, and anyone wanting to reach the cutting edge has a lot of catching up to do. But, later in my career, my research led me to Grothendieck's work.

Had I taken a different path, I might have come to grips with his work through string theory. String theorists postulate that besides the visible dimensions of space and time—three of space and one of time—there are extra dimensions of space curled up too small to see. In some of their theories, these extra dimensions form a variety. So, string theorists are easily pulled into sophisticated questions about algebraic geometry. And this, in turn, pulls them toward Grothendieck.

Indeed, some of the best advertisements for string theory are not successful predictions of experimental results—it has made absolutely none of these—but rather, its ability to solve problems within pure mathematics, including algebraic geometry. For example, string theory is shockingly good at counting how many curves of different kinds you can draw on certain varieties. Thus, we now see string theorists talking with algebraic geometers, each able to surprise the other with their insights.

My own interest in Grothendieck's work had a different source. I have always had serious doubts about string theory, and counting curves on varieties is the last thing I would ever try: Like rock climbing, it is exciting to watch but too scary to actually attempt myself. It turns out that Grothendieck's ideas are so general and powerful that they spill out beyond algebraic geometry into many other subjects. In particular, his 600-page unpublished manuscript *Pursuing Stacks*, written in 1983, made a big impression on me. In it, he argued that topology—very loosely, the theory of what space can be shaped like, if we do not care about bending or stretching it, just what kind of holes it has—can be completely reduced to algebra!

At first, this idea may sound just like algebraic geometry, where we use algebra to describe geometrical shapes, like curves or higher dimensional varieties. But "algebraic topology" winds up having a very different flavor because in topology we do not restrict our shapes to be described by polynomial equations. Instead of dealing with beautiful gems, we are dealing with floppy, flexible blobs—so the kind of algebra we need is different.

Algebraic topology is a beautiful subject that was around long before Grothendieck—but he was one of the first to seriously propose a method to reduce *all* topology to algebra. Thanks to my work on physics, his proposal was tremendously exciting when I came across it. Here is why: At the time, I had taken up the challenge of trying to unify our two best theories of physics: quantum physics, which describes all the forces except gravity, and general relativity, which describes gravity. It seems that until we do this, our understanding of the fundamental laws of physics is doomed to be incomplete. But it is devilishly difficult. One reason is that quantum physics is based on algebra, whereas general relativity involves a lot of topology. But that reason suggests an avenue of attack: If we can figure out how to reduce topology to algebra, it might help us formulate a theory of quantum gravity.

My physics colleagues will let out a howl here and complain that I am oversimplifying. Yes, I am oversimplifying: There is more to quantum physics than mere algebra, and more to general relativity than mere topology. Nonetheless, the possible benefits to physics of reducing topology to algebra are what made me so excited about Grothendieck's work.

So, starting in the 1990s, I tried to understand the powerful abstract concepts that Grothendieck had invented—and by now I have partially succeeded. Some mathematicians find these concepts to be the hard part of algebraic geometry. They now seem like the easy part to me. The hard part, for me, is not these abstract concepts but the nitty-gritty details. First, there is all the material in those texts that Hartshorne takes as prerequisites: "the books by Atiyah–Macdonald, Matsumura, and Zariski–Samuel"—in short, piles and piles of algebra. But there is also a lot more.

So, while I now have *some* of what it takes to read Hartshorne, until recently I was too intimidated to learn it. A student of physics once asked a famous expert how much mathematics a physicist needs to know. The expert replied: "More." Indeed, the job of learning mathematics is never done, so I focus on the things that seem most important and/or fun. Until last year, algebraic geometry never rose to the top of the list.

What changed? I realized that algebraic geometry is connected to the relation between classical and quantum physics. Classical physics is the physics of Newton, where we imagine that we can measure everything with complete precision, at least in principle. Quantum physics is the physics of Schrödinger and Heisenberg, governed by the uncertainty principle: If we measure some aspects of a physical system with complete precision, others must remain undetermined.

For example, any spinning object has an "angular momentum." In classical mechanics, we visualize this as an arrow pointing along the axis of rotation, whose length is proportional to how fast the object is spinning. And in classical mechanics, we assume that we can measure this arrow precisely. In quantum mechanics—a more accurate description of reality—this turns out not to be true. For example, if we know how far this arrow points in the x direction, we cannot know how far it points in the y direction. This uncertainty is too small to be noticeable for a spinning basketball, but for an electron it is important: Physicists had only a rough understanding of electrons until they took this problem into account.

Physicists often want to "quantize" classical physics problems. That is, they start with the classical description of some physical system, and they want to figure out the quantum description. There is no fully general and completely systematic procedure for doing this. This should not be surprising: The two world views are so different. However, there *are* useful recipes for quantization. The most systematic ones apply to a very limited selection of physics problems.

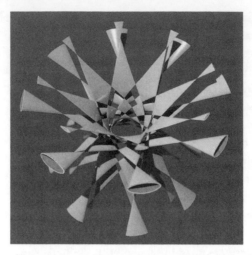

Variety: The Endrass octic is a beautiful, highly symmetrical example of a variety: a shape described by polynomial equations. Algebraic geometry began as the study of such shapes. Image by Abdelaziz Nait Merzouk. See also color insert.

For example, sometimes in classical physics we can describe a system by a point in a *variety*. This is not something one generally expects, but it happens in plenty of important cases. For example, consider a spinning object: If we fix how long its angular momentum arrow is, the arrow can still point in any direction, so its tip must lie on a sphere. Thus, we can describe a spinning object by a point on a sphere. And this sphere is actually a variety, the "Riemann sphere,"[1] named after Bernhard Riemann, one of the greatest algebraic geometers of the 1800s.

When a classical physics problem is described by a variety, some magic happens. The process of quantization becomes completely systematic—and surprisingly simple. There is even kind of a reverse process, which one might call "classicization," that lets you turn the quantum description back into a classical description. The classical and quantum approaches to physics become tightly linked, and one can take ideas from either approach and see what they say about the other one. For example, each point on the variety describes not only a state of the classical system (in our example, a definite direction for the angular momentum), but also a state of the corresponding quantum system—even though the latter is governed by Heisenberg's uncertainty principle. The quantum state is the "best quantum approximation" to the classical state. Even better, in this situation, many of the basic theorems about algebraic geometry can be seen as facts about quantization. Since quantization is something I have been thinking about for a long time, this makes me very happy.

Richard Feynman once said that, for him to make progress on a tough physics problem, he needed to have some sort of special angle on it:

> I have to think that I have some kind of inside track on this problem. That is, I have some sort of talent that the other guys aren't using, or some way of looking, and they are being foolish not to notice this wonderful way to look at it. I have to think I have a little better chance than the other guys, for some reason. I know in my heart that it is likely that the reason is false, and likely the particular attitude I'm taking with it was thought of by others. I don't care; I fool myself into thinking I have an extra chance.

This may be what I had been missing on algebraic geometry until now. Of course, algebraic geometry is not just a problem to be solved, it is a body of knowledge—but it is such a large, intimidating body of knowledge that I did not dare tackle it until I got an inside track. Now I can read Hartshorne, translate some of the results into facts about physics, and feel I have a chance at understanding this stuff.[2] And it is a great feeling.

Notes

1. https://en.wikipedia.org/wiki/Riemann_sphere.

2. John Baez, From Classical to Quantum and back, available at http://math.ucr.edu/home/baez/gq/.

Decades-Old Computer Science Conjecture Solved in Two Pages

ERICA KLARREICH

A paper posted online in July 2019 has settled a nearly 30-year-old conjecture about the structure of the fundamental building blocks of computer circuits.[1] This "sensitivity" conjecture has stumped many of the most prominent computer scientists over the years, yet the new proof is so simple that one researcher summed it up in a single tweet.[2]

"This conjecture has stood as one of the most frustrating and embarrassing open problems in all of combinatorics and theoretical computer science," wrote Scott Aaronson[3] of the University of Texas, Austin, in a blog post.[4] "The list of people who tried to solve it and failed is like a who's who of discrete math and theoretical computer science," he added in an email.

The conjecture concerns Boolean functions, rules for transforming a string of input bits (0s and 1s) into a single output bit. One such rule is to output a 1 provided any of the input bits is 1, and a 0 otherwise; another rule is to output a 0 if the string has an even number of 1s, and a 1 otherwise. Every computer circuit is some combination of Boolean functions, making them "the bricks and mortar of whatever you're doing in computer science," said Rocco Servedio[5] of Columbia University.

Over the years, computer scientists have developed many ways to measure the complexity of a given Boolean function. Each measure captures a different aspect of how the information in the input string determines the output bit. For instance, the "sensitivity" of a Boolean function tracks, roughly speaking, the likelihood that flipping a single input bit will alter the output bit. And "query complexity" calculates how many input bits you have to ask about before you can be sure of the output.

Each measure provides a unique window into the structure of the Boolean function. Yet computer scientists have found that nearly all these measures fit into a unified framework, so that the value of any one of them is a rough gauge for the value of the others. Only one complexity measure did not seem to fit in: sensitivity.

In 1992, Noam Nisan[6] of the Hebrew University of Jerusalem and Mario Szegedy[7], now of Rutgers University, conjectured that sensitivity does indeed fit into this framework.[8] But no one could prove it. "This, I would say, probably was the outstanding open question in the study of Boolean functions," Servedio said.

"People wrote long, complicated papers trying to make the tiniest progress," said Ryan O'Donnell[9] of Carnegie Mellon University.

Now Hao Huang,[10] a mathematician at Emory University, has proved the sensitivity conjecture with an ingenious but elementary two-page argument about the combinatorics of points on cubes. "It is just beautiful, like a precious pearl," wrote Claire Mathieu,[11] of the French National Center for Scientific Research, during a Skype interview.

Aaronson and O'Donnell both called Huang's paper the "book" proof of the sensitivity conjecture, referring to Paul Erdős's notion of a celestial book in which God writes the perfect proof of every theorem.[12] "I find it hard to imagine that even God knows how to prove the Sensitivity Conjecture in any simpler way than this," Aaronson wrote.[13]

A Sensitive Matter

Imagine, Mathieu said, that you are filling out a series of yes/no questions on a bank loan application. When you are done, the banker will score your results and tell you whether you qualify for a loan. This process is a Boolean function: Your answers are the input bits, and the banker's decision is the output bit.

If your application gets denied, you might wonder whether you could have changed the outcome by lying on a single question—perhaps, by claiming that you earn more than $50,000 when you really do not. If that lie would have flipped the outcome, computer scientists say that the Boolean function is "sensitive" to the value of that particular bit. If, say, there are seven different lies you could have told that would have each separately flipped the outcome, then for your loan profile, the sensitivity of the Boolean function is seven.

Computer scientists define the overall sensitivity of the Boolean function as the biggest sensitivity value when looking at all the different possible loan profiles. In some sense, this measure calculates how many of the questions are truly important in the most borderline cases—the applications that could most easily have swung the other way if they had been ever so slightly different.

Sensitivity is usually one of the easiest complexity measures to compute, but it is far from the only illuminating measure. For instance, instead of handing you a paper application, the banker could have interviewed you, starting with a single question and then using your answer to determine what question to ask next. The largest number of questions the banker would ever need to ask before reaching a decision is the Boolean function's query complexity.

This measure arises in a host of settings—for instance, a doctor might want to send a patient for as few tests as possible before reaching a diagnosis, or a machine learning expert might want an algorithm to examine as few features of an object as possible before classifying it. "In a lot of situations—diagnostic situations or learning situations—you're really happy if the underlying rule . . . has low query complexity," O'Donnell said.

Other measures involve looking for the simplest way to write the Boolean function as a mathematical expression, or calculating how many answers the banker would have to show a boss to prove they had made the right loan decision. There is even a quantum physics version of query complexity in which the banker can ask a "superposition" of several questions at the same time. Figuring out how this measure relates to other complexity measures has helped researchers understand the limitations of quantum algorithms.[14]

With the single exception of sensitivity, computer scientists proved that all these measures are closely linked. Specifically, they have a polynomial relationship to each other—for example, one measure might be roughly the square or cube or square root of another. Only sensitivity stubbornly refused to fit into this neat characterization. Many researchers suspected that it did indeed belong, but they could not prove that there were no strange Boolean functions out there whose sensitivity had an exponential rather than polynomial relationship to the other measures, which in this setting would mean that the sensitivity measure is vastly smaller than the other measures.

"This question was a thorn in people's sides for 30 years," Aaronson said.

Boolean Sensitivities

To visualize how sensitive a computer circuit is to bit-flip errors, we can represent its n input bits as the coordinates of a corner of an n-dimensional cube and color the corner blue if the circuit outputs 1 and red if it outputs 0.

Circuits and Bit-Flips ─────────────────────

| OR | Output **1** if **any** input bit is 1. Output **0** if **all** input bits are 0. | AND | Output **1** if **all** input bits are 1. Output **0** if **any** input bit is 0. |

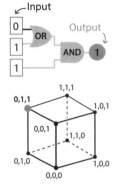

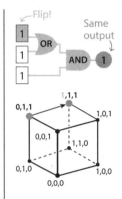

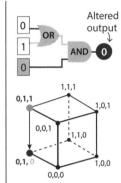

The output of this simple Boolean function with the input string 011 can be represented as a blue dot at the (0,1,1) corner of this 3D cube.

If you flip the first bit, you move to the blue (1,1,1) corner of the cube. The function is not sensitive to this bit flip.

If instead you flip the third bit, you move to the red (0,1,0) corner of the cube. The function is sensitive to this bit flip.

Measuring Sensitivity ─────────────────────

Once every corner of the cube has been colored according to our Boolean function, the number of sensitive bits for a given input string is captured by the number of connections between its associated corner and corners of the other color. A circuit's overall sensitivity is defined as the largest number of sensitive bits in any input string, so this Boolean function's **sensitivity is 2.**

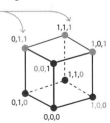

Infographic by Lucy Reading-Ikkanda (see also color insert)

Cornering the Solution

Huang heard about the sensitivity conjecture in late 2012, over lunch with the mathematician Michael Saks[15] at the Institute for Advanced Study, where Huang was a postdoctoral fellow. He was immediately taken with the conjecture's simplicity and elegance. "Starting from that moment, I became really obsessed with thinking about it," he said.

Huang added the sensitivity conjecture to a "secret list" of problems he was interested in, and whenever he learned about a new mathematical tool, he considered whether it might help. "Every time after I'd publish a new paper, I would always go back to this problem," he said. "Of course, I would give up after a certain amount of time and work on some more realistic problem."

Huang knew, as did the broader research community, that the sensitivity conjecture could be settled if mathematicians could prove an easily stated conjecture about collections of points on cubes of different dimensions. There is a natural way to go from a string of n 0s and 1s to a point on an n-dimensional cube: Simply use the n bits as the coordinates of the point.

For instance, the four two-bit strings—00, 01, 10, and 11—correspond to the four corners of a square in the two-dimensional plane: (0,0), (0,1), (1,0) and (1,1). Likewise, the eight three-bit strings correspond to the eight corners of a three-dimensional cube, and so on in higher dimensions. A Boolean function, in turn, can be thought of as a rule for coloring these corners with two different colors (say, red for 0 and blue for 1).

In 1992, Craig Gotsman,[16] now of the New Jersey Institute of Technology, and Nati Linial[17] of Hebrew University figured out that proving the sensitivity conjecture can be reduced to answering a simple question about cubes of different dimensions: If you choose any collection of more than half the corners of a cube and color them red, is there always some red point that is connected to many other red points?[18] (Here, by "connected," we mean that the two points share one of the outer edges of the cube, as opposed to being across a diagonal.)

If your collection contains exactly half the corners of the cube, it is possible that none of them will be connected. For example, among the eight corners of the three-dimensional cube, the four points (0,0,0), (1,1,0), (1,0,1) and (0,1,1) all sit across diagonals from one another. But as soon as more than half the points in a cube of any dimension are colored red, some connections between red points must pop up. The question is: How are these connections distributed? Will there be at least one highly connected point?

In 2013, Huang started thinking that the best route to understanding this question might be through the standard method of representing a network with a matrix that tracks which points are connected and then examining a set of numbers called the matrix's eigenvalues. For five years, he kept revisiting this idea, without success. "But at least thinking about it [helped] me quickly fall asleep many nights," he commented on Aaronson's blog post.[19]

Then in 2018, it occurred to Huang to use a 200-year-old piece of mathematics called the Cauchy interlace theorem, which relates a matrix's eigenvalues to those of a submatrix, making it potentially the perfect tool to study the relationship between a cube and a subset of its corners. Huang decided to request a grant from the National Science Foundation to explore this idea further.

Then in June 2019, as he sat in a Madrid hotel writing his grant proposal, he suddenly realized that he could push this approach all the way to fruition simply by switching the signs of some of the numbers in his matrix. In this way, he was able to prove that in any collection of more than half the points in an n-dimensional cube, there will be some point that is connected to at least $\sqrt{n}$ of the other points—and the sensitivity conjecture instantly followed from this result.

When Huang's paper landed in Mathieu's inbox, her first reaction was "uh-oh," she said. "When a problem has been around 30 years and everybody has heard about it, probably the proof is either very long and tedious and complicated, or it's very deep." She opened the paper expecting to understand nothing.

But the proof was simple enough for Mathieu and many other researchers to digest in one sitting. "I expect that this fall it will be taught—in a single lecture—in every master's-level combinatorics course," she messaged over Skype.

Huang's result is even stronger than necessary to prove the sensitivity conjecture, and this power should yield new insights about complexity measures. "It adds to our toolkit for maybe trying to answer other questions in the analysis of Boolean functions," Servedio said.

Most importantly, though, Huang's result lays to rest nagging worries about whether sensitivity might be some strange outlier in the world of complexity measures, Servedio said. "I think a lot of people slept easier that night, after hearing about this."

Notes

1. https://arxiv.org/abs/1907.00847.
2. Ryan O'Donnell
   @BooleanAnalysis
   Ex.1: ∃edge-signing of n-cube with 2^{n-1} eigs each of $+/-sqrt(n)$
   Interlacing=>Any induced subgraph with $>2^{n-1}$ vtcs has max eig $>= sqrt(n)$
   Ex.2: In subgraph, max eig $<=$ max valency, even with signs
   Hence [GL92] the Sensitivity Conj, $s(f) >= sqrt(deg(f))$
3. https://www.cs.utexas.edu/people/faculty-researchers/scott-aaronson.
4. https://www.scottaaronson.com/blog/?p=4229.
5. http://www.cs.columbia.edu/~rocco/.
6. https://www.cse.huji.ac.il/~noam/.
7. https://www.cs.rutgers.edu/~szegedy/.
8. https://dl.acm.org/doi/10.1145/129712.129757.
9. http://www.cs.cmu.edu/~odonnell/.
10. http://www.math.emory.edu/people/faculty/individual.php?NUM=397.
11. https://www.di.ens.fr/ClaireMathieu.html.
12. https://www.quantamagazine.org/gunter-ziegler-and-martin-aigner-seek-gods-perfect-math-proofs-20180319/.
13. https://www.scottaaronson.com/blog/?p=4229.
14. https://www.quantamagazine.org/quantum-computers-struggle-against-classical-algorithms-20180201/.
15. https://sites.math.rutgers.edu/~saks/.
16. http://www.cs.technion.ac.il/~gotsman/.
17. https://www.cse.huji.ac.il/~nati/.
18. https://dl.acm.org/doi/10.1016/0097-3165%2892%2990060-8.
19. https://www.scottaaronson.com/blog/?p=4229#comment-1813116.

The Three-Body Problem

RICHARD MONTGOMERY

By the spring of 2014, I had largely given up on the three-body problem. Out of ideas, I began programming on my laptop to generate and search through approximate solutions.

These attempts would never solve my problem outright, but they might garner evidence toward an answer. My lack of programming expertise and resulting impatience slowed the process, making it an unpleasant experience for a pencil-and-paper mathematician like myself. I sought out my old friend Carles Simó, a professor at the University of Barcelona, to convince him to aid me in my clunky search.

That fall I traveled to Spain to meet with Simó, who had a reputation as one of the most inventive and careful numerical analysts working in celestial mechanics. He is also a direct man who does not waste time or mince words. My first afternoon in his office, after I had explained my question, he looked at me with piercing eyes and asked, "Richard, why do you care?"

The answer goes back to the origins of the three-body problem. Isaac Newton originally posed and solved the two-body problem (Figure 1) when he published his *Principia* in 1687. He asked, "How will two masses move in space if the only force on them is their mutual gravitational attraction?" Newton framed the question as a problem of solving a system of differential equations—equations that dictate an object's future motion from its present position and velocity. He completely solved his equations for two bodies. The solutions, also called orbits, have each object moving on a conic—a circle, ellipse, parabola, or hyperbola. In finding all the possible orbits, Newton derived Johannes Kepler's laws of planetary motion, empirical laws Kepler published in 1609 that synthesized decades of astronomical observations by his late employer, Tycho Brahe. Kepler's first law says that

Mass of sun much greater than planet:
sun's elliptical orbit is tiny.

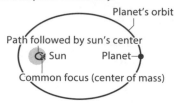

Two equal masses in elliptical orbits

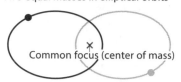

FIGURE 1. Two examples of two-body orbits.

each planet (or comet) moves on a conic with the sun as its focus. In Newton's solutions, however, the two bodies—the sun and a planet—move on two separate conics. These conics share one focus, which is the center of mass of the two bodies. The sun is more massive than any planet, so much so that the center of the mass of the sun-planet system is inside the sun itself, very close to the sun's center of mass, with the sun's center of mass barely wobbling about the common center on a tiny elliptical path.

In place of the two masses, put three, and you have the three-body problem. Like its predecessor, its orbits are solutions to a system of differential equations. Unlike its predecessor, however, it is difficult to impossible to find explicit formulas for the orbits. To this day, despite modern computers and centuries of work by some of the best physicists and mathematicians, we only have explicit formulas for five families of orbits, three found by Leonhard Euler (in 1767) and two by Joseph-Louis Lagrange (in 1772) (Figure 2). In 1890, Henri Poincaré discovered chaotic dynamics within the three-body problem, a finding that implies we can never know all the solutions to the problem at a level of detail remotely approaching Newton's complete solution to the two-body problem. Yet through a process called numerical integration, done efficiently on a computer, we can nonetheless generate finite segments of approximate orbits, a process essential to the planning of space missions. By extending the runtime of the computer, we can make the approximations as accurate as we want.

One of Euler's solutions
Three equal masses with one at the center: bodies are always collinear.

One of Lagrange's solutions
The bodies form an equilateral triangle at all times.

FIGURE 2. Two examples of three-body orbits. (See also color insert.)

Eclipses

Simó's words had knocked the breath out of me. "Of course, I care," I thought. "I have been working on this problem for nearly two decades!" In fact, I had been focusing on a particular question within the problem that interested me:

> Is every periodic eclipse sequence the eclipse sequence of some periodic solution to the planar three-body problem?

Let me explain. Imagine three bodies—think of them as stars or planets—moving about on a plane, pulling at one another with gravity. Number the bodies one, two, and three. From time to time, all three will align in a single, straight line (Figure 3). Think of these moments as eclipses. (Technically, this "eclipse" is called a *syzygy*, an unbeatable word to use in hangman.) As time passes, record each eclipse as it occurs, labeling it one, two, or three, for whichever star is in the middle. In this way, we get a list of ones, twos, and threes called the eclipse sequence.

For example, in a simplified version of our sun-Earth-moon system, the moon (which we will label body "3") makes a circle around Earth (body "2") every month, while Earth makes a circle around the sun (body "1") once a year. This movement is repetitive, so it will give us a periodic eclipse sequence. Specifically: 2, 3, 2, 3, 2, 3, 2, 3, 2, 3, 2, 3, 2, 3, 2, 3, 2, 3, 2, 3, 2, 3, 2, 3. There is no 1 in the sequence because

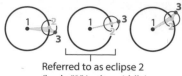

Referred to as eclipse 2
(body "2" in the middle)   FIGURE 3. Eclipse moment.

the sun never lands between Earth and the moon. In one year, the list is 24 numbers long, with a 2, 3 for each of the 12 months of the year.

There is no reason that the eclipse sequence of a solution must repeat itself. It might go on forever with no discernible pattern. If, however, the solution exactly repeats itself after some period of time, like the Earth-moon-sun system after a year, then the sequence repeats: The same 24 numbers of the Earth-moon-sun system replay each year. So, returning to my question: Is every periodic eclipse sequence the eclipse sequence of some periodic solution to the planar three-body problem? I suspected the answer was yes, but I could not prove it.

Holey Objects

To justify the importance of my question, I reminded Simó of a basic fact tying together three branches of mathematics: topology, sometimes called rubber sheet geometry; Riemannian geometry, the study of curved surfaces; and dynamics, the study of how things move. Imagine a bug walking along a curved surface shaped like the "wormhole surface," also called a catenoid (Figure 4). The bug's job is to find the shortest circuit going once around the hole. As far as topology is concerned, the wormhole surface is the same as the *x-y* plane with a single hole punctured in it. Indeed, imagine a hole punctured into a flexible rubber sheet. By pushing the hole downward and stretching it outward, you can make the wormhole surface (Figure 5). If the hole has been sufficiently flared outward, then not only does this shortest circuit exist, but it satisfies a differential equation very much like the three-body equations. In this way, our bug has found a periodic solution to an interesting differential equation.

In the three-body problem, the role of the wormhole surface is played by something called configuration space—a space whose points encode the locations of all three bodies simultaneously, so that a curve

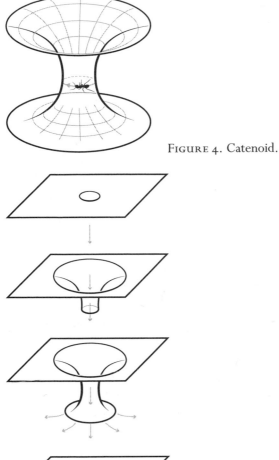

FIGURE 4. Catenoid.

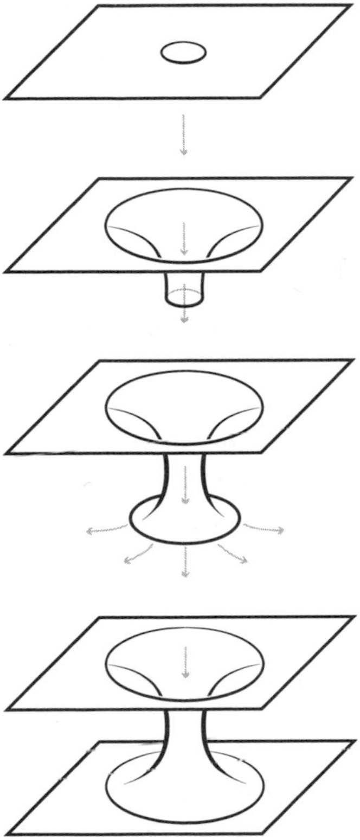

FIGURE 5. Rubber sheet geometry.

in configuration space specifies the motions of each of the three bodies. By insisting that our bodies do not collide with one another, we pierce holes in this configuration space. As we will see, as far as topology, or rubber sheet geometry, is concerned, the resulting collision-free configuration space is the same as an *x-y* plane with two holes punctured in

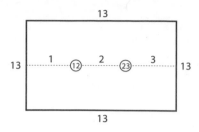

FIGURE 6. Collision-free configuration space.

it (Figure 6). We label the holes as "12," meaning bodies 1 and 2 have collided, and "23," meaning that 2 and 3 have collided, and place the holes on the *x*-axis. We also place a third hole at infinity and label it "13" to represent bodies 1 and 3 colliding. These holes break the *x*-axis into three segments labeled 1, 2, and 3. A curve in this twice-punctured plane represents a motion of all three bodies—which is to say, a potential solution to the three-body problem. When the curve cuts across segment 1, it means an eclipse of type 1 has occurred, and likewise for cutting across segment 2 or 3. In this way, an eclipse sequence represents a way of winding around our collision holes.

Now, our bug was trying to minimize the length of its path as it circled once around the wormhole. To get the correct analogy between the bug's problem and the three-body problem, we must replace the length of a path by a quantity called the *action of a path*. (The action is a kind of average of the instantaneous kinetic energy minus the potential energy of the motion represented by the path.) A centuries-old theorem from mechanics states that any curve in configuration space that minimizes the action must be a solution to Newton's three-body problem. We can thus try to solve our eclipse sequence problem by searching, among all closed paths that produce a fixed eclipse sequence, for those closed paths that minimize the action.

This strategy—seeking to minimize the action in configuration space for loops that have a particular eclipse sequence—had preoccupied me for most of 17 years and led to many nice results. For instance, in 2000, Alain Chenciner of Paris Diderot University and I rediscovered what seems to be the first known periodic solution to the three-body problem with zero angular momentum. It was a figure-eight-shaped solution (Figure 7) first found by Cris Moore of the Santa Fe Institute in 1993. In this case, three equal masses chase one another around a figure-eight shape on the plane. Its eclipse sequence is 123123,

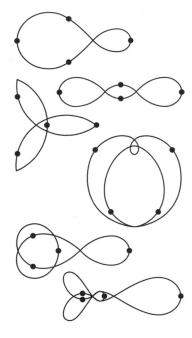

FIGURE 7. Figure eight solution.

FIGURE 8. New orbits.

repeating forever. Our work popularized the figure eight and gave it a rigorous existence proof. It also led to an explosion of discoveries of many new orbits for the equal-mass N-body problem, orbits christened "choreographies" by Simó (Figure 8), who discovered hundreds of these new families of orbits. Our figure-eight orbit even made it into the best-selling Chinese science fiction novel by Liu Cixin, whose English translation was entitled *The Three-Body Problem*.

The morning after I shared my ponderings with Simó, he said something that affected me deeply. "Richard, if what you think about your question is true, then there must be a dynamical mechanism." In other words, if I was right that the answer to my question was yes, then there must be something about how these bodies moved that made it so.

Those few words made me question my convictions and led me to abandon my 17-year-long attempt to answer my question by minimizing

the action of paths. What dynamical mechanisms in this problem did I
even understand? I wondered. I could think of two, only one of which
held out hope. This mechanism, related to the chaos discovered by Poincaré, led me to reflect on old work of a recent collaborator of mine, Rick
Moeckel of the University of Minnesota. In the 1980s, he had shown
how curves called *hyperbolic tangles*, born from triple collisions in the
three-body problem, can lead to astounding results. As I reread his old
papers, it seemed to me that Moeckel had the key to my problem. I got
in touch with him, and within a few days Moeckel and I had answered
my question! Well, almost. We had answered a question infinitely close.

The Shape Sphere

Understanding Moeckel's dynamical mechanism, in conjunction with
the relationship between the three-body configuration space and the
plane with two holes described above, requires thinking about an object called the *shape sphere*. As the three bodies move around in the
plane, at each instant they form the three vertices of a triangle. Instead
of keeping track of the position of each vertex, let us keep track of only
the overall shape of the triangle. The result is a curve on the shape
sphere, a sphere whose points represent "shapes" of triangles.

What is a "shape"? Two figures in the plane have the same shape if
we can change one figure into the other by translating, rotating, or
scaling it (Figure 9). The operation of passing from the usual three-
body configuration space—which is to say, from the knowledge of the
locations of all three vertices of a triangle—to a point in the shape
sphere, is a process of forgetting—forgetting the size of the triangle,
the location of its center of mass, and the orientation of the triangle in
the plane. That the shape sphere is two-dimensional is easy to understand from high school geometry: We know the shape of a triangle if
we know all three of its angles, but because the sum of the three angles
is always 180 degrees, we really only need two of the three angles—
hence, two numbers are sufficient to describe the shape of a triangle.
That the shape sphere is actually a sphere is harder to understand and
requires that we allow triangles to degenerate, which is to say, we allow
"triangles" consisting of three vertices that all lie on the same line to
be called triangles. These so-called degenerate zero-area triangles form
the equator of the shape sphere: they are the eclipses!

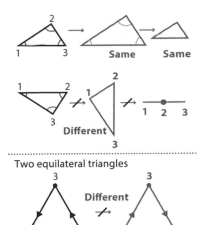

Two equilateral triangles

Lagrange point 4 **Lagrange point 5** FIGURE 9. Shapes.

The area of a triangle, divided by its size (r) squared, is its distance to the equator. The north and south poles of the sphere represent those triangles of maximum possible area and are the two equilateral triangle shapes. But why are there two equilateral shapes? These two equilateral triangle shapes differ by the cyclic order of their vertices. There is no way to turn one of these equilateral triangles into the other by a rotation, translation, or scaling of the plane: they represent different shapes. Yet the operation of reflecting about a line (any line) in the plane will turn one equilateral triangle shape into the other one. This reflection operation acts on all triangles, and so on the shape sphere itself, where it acts by reflection about the equator, keeping the points of the equator (degenerate triangles) fixed while interchanging the north and south hemispheres.

Included among the degenerate triangles are the binary collisions: those "triangles" for which two of the three vertices lie on top of each other. There are exactly three of these binary collision triangles, labeled "12," "23," and "13," according to which two vertices lie on top of each other.

I can now explain how the shape sphere shows us that the three-body configuration space is topologically the same as the usual x-y plane minus two points. We have to know that the sphere minus a single point is topologically the same object as the usual x-y plane. One way to

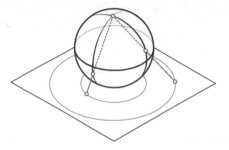

FIGURE 10. Stereographic projection.

see this fact about the sphere is to use stereographic projection, which maps the sphere with a single point removed (the "light source") onto the usual x-y plane (Figure 10). As a point on the sphere tends toward the light source, its image point on the x-y plane moves out to infinity, so we can also say that the plane with a point at infinity added is topologically equal to the sphere. Take the light source to be the 13 binary collision point of the shape sphere, so that the point at infinity of the x-y plane corresponds to the 13 collision point. Orient the sphere so that its equatorial plane intersects the x-axis of the x-y plane. Then stereographic projection maps the equator of degenerate triangles to the x-axis of the plane, and the other two binary collision points are mapped to two points on this x-axis. In this way, we arrive at exactly the picture described earlier.

The three binary collision points form three special points on the shape sphere (Figure 11). Besides these three, there are additional special points on the shape sphere called *central configurations*. These five central configurations correspond to the five families of solutions discovered by Euler and Lagrange. Their solutions are the only three-body solutions for which the shape of the triangle does not change as the triangle evolves! In the Lagrange solutions, the triangle remains equilateral at each instant; there are two Lagrange configurations, as we have seen, and they form the north and south poles of the shape sphere. We label them "Lagrange point 4" and "Lagrange point 5." The remaining three central configurations are the Euler configurations, labeled "Euler point 1," "Euler point 2," and "Euler point 3." They are collinear (all in a line), degenerate configurations, so they lie on the equator of the shape sphere. They are positioned on the equator between the three binary collision points (Figure 12). (Their spacing along the equator depends on the mass ratios between the three masses of the bodies.)

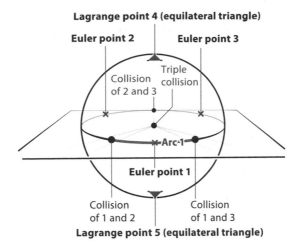

FIGURE 11. The shape sphere.

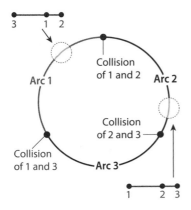

FIGURE 12. Binary collision points and eclipses.

Euler point 1, for example, lies on the equatorial arc marked 1, so is a collinear shape in which body 1 lies between bodies 2 and 3. (Often all five central configuration points are called Lagrange points, with the Euler points labeled "L1," "L2," and "L3.")

One can understand the central configuration solutions by dropping three bodies, by which I mean, by letting the three bodies go from rest, with no initial velocity. Typically when one does this, all kinds of crazy things happen: close binary collisions, wild dances, and perhaps the escape of one body to infinity. But if one drops the three bodies when they are arranged in one of the five central configuration shapes, then the triangle they form simply shrinks to a point, remaining in precisely the same shape as it started, with the three masses uniformly pulling on one another until the solution ends in a simultaneous triple collision.

The Five Roads to Triple Collision

Triple collision is an essential singularity within the three-body problem, something like a big bang at the center of the problem, and it is the source of much of its chaos and difficulty. In the early 1900s, Finnish mathematician Karl Sundman proved that the five central configurations, as represented by the dropped solutions just described, are the only roads to triple collision. What this means is that any solution that ends in triple collision must approach it in a manner very close to one of these five dropped central configuration solutions, and as it gets closer and closer to triple collision, the shape of the solution must approach one of the five central configuration shapes.

Sundman's work was a complicated feat of algebra and analysis. Then, in the year I graduated from high school (completely oblivious to the three-body problem), American mathematician Richard McGehee invented his so-called blow-up method, which allowed us to understand Sundman's work pictorially and to study dynamics near triple collision in much greater detail. Let r denote the distance to triple collision—a measure of the overall size of a triangle. As r approaches zero, Newton's equations become very badly behaved, with many terms going to infinity. McGehee found a change of configuration space variables and of time that slows down the rate of approach to triple collision and turns the triple collision point, which is $r = 0$, into an entire collection of points: the collision manifold. Surprise! The collision manifold is essentially the shape sphere. McGehee's method extended Newton's equations, originally only valid for r greater than zero, to a system of differential equations that makes sense when $r = 0$.

Newton's equations have no equilibrium points, meaning there are no configurations of the three bodies that stand still: Three stars, all attracting one another, cannot just sit there in space without moving. But when Newton's equations are extended to the collision manifold, equilibrium points appear. There are exactly 10 of them, a pair for each of the five central configuration points on the shape sphere. One element of a pair represents the end result of the corresponding dropped central configuration in its approach to triple collision. Newton's equations stay the same even if we run time backward, so we can run any solution in reverse and get another solution. When we run a dropped central configuration solution backward, we get a solution that explodes out of triple collision, reaching its maximum size at the dropped configuration. The other element of the pair represents the initial starting point of this "exploding" solution. Together, these two central configuration solutions—collision and ejection—fit smoothly and form a single ejection-collision solution that leaves the ejection equilibrium point at $r = 0$, enters into the r greater than zero region where it achieves a maximum size, and then shrinks back to end up on the triple collision manifold at the collision equilibrium point there. This complete solution connects one element of an equilibrium pair to the other.

By creating these equilibrium points associated with central configurations, buried deep inside the three-body problem, McGehee gave Moeckel a key that enabled him to apply recently established results from modern dynamical systems—results unavailable to Newton, Lagrange, or Sundman—to make some interesting headway on the three-body problem.

Moeckel's Walk

In Moeckel's papers, I saw a picture of a graph with five vertices labeled by the central configurations and joined together by edges (Figure 13).

A walk on a graph is a possible circuit through its vertices, traveling the edges from vertex to vertex. Moeckel proved that any possible walk you can take on his graph corresponds to a solution to the three-body problem that comes close for some time to the central configuration

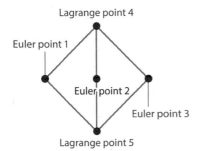

FIGURE 13. Moeckel's graph.

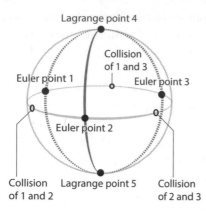

Collision   Lagrange point 5   Collision   FIGURE 14. Moeckel's graph,
of 1 and 2   of 2 and 3   embedded.

solution labeled by the corresponding vertex. For example, the walk
E1 L4 E2 L5 corresponds to a solution which starts close to the Euler
ejection-collision solution associated with the Euler point 1, then
comes close to triple collision almost along the Lagrange L4 central
configuration solution, but before total triple collision is achieved the
three bodies shoot out along one of the five "roads" very near to the
Euler point 2 central configuration solution. Then, finally, as this Euler
solution collapses back toward triple collision, the solution spins out
into a Lagrange L5 equilateral shape. Moreover, if we repeat this same
walk, making it periodic, the solution following it will be periodic.

Soon after Simó told me there had to be a dynamical mechanism,
I realized that Moeckel's graph embedded into the shape sphere (Figure 14). The important thing about this embedded graph is that it carries all of the topology of the sphere with its three binary collision
holes. Indeed, we can deform the thrice-punctured sphere onto the
graph and in so doing turn any loop in the punctured sphere to a walk
on the graph. To see this deformation, imagine the sphere as the surface
of a balloon. Make three pin pricks in it, one at each binary collision
hole. The balloon is made of very flexible material, so we can stretch
out our three pinpricks, enlarging them until the edges of the three
holes almost touch each other and the remaining material forms a ribbon hugging close to the embedded graph. In the process of making
this deformation, any closed loop in the thrice-punctured sphere gets
deformed into a closed loop in this ribbon structure and, from there, to
a walk on Moeckel's embedded graph.

To turn this picture into a theorem about solutions, I needed to prove that if I project the solutions guaranteed by Moeckel's theorem onto the shape sphere, then they never stray far from this embedded graph. If they did, they could wind around the binary collisions or even hit one, killing or adding some topologically significant loops and so changing the eclipse sequence. I emailed Moeckel to ask for help. He wrote back, "You mean you're going to force me to read papers I wrote over 20 years ago?" Nevertheless, he dove back into his old research and proved that the projections of the solutions he had encoded symbolically all those years ago never did stray far from the embedded graph. My question was answered—almost.

To make his proof work, Moeckel needed a tiny bit of angular momentum. (Angular momentum, in this context, is a measure of the total amount of "spin" of a system and is constant for each solution.) But for those 17 years before my conversation with Simó, I had insisted on solutions having zero angular momentum. This insistence arose because solutions that minimize action among all curves that have a given eclipse sequence must have zero angular momentum. On the other hand, Moeckel needed a small bit of angular momentum to get solutions traveling along the edges of his graph. The symbol for a tiny positive quantity in mathematical analysis is an epsilon. We needed an epsilon of angular momentum.

There was another catch to Moeckel's results: His solutions, when they cross the equator of the shape sphere near the Euler points E1, E2, and E3, oscillate back and forth there across the equator before traveling up to the north or south pole as they go in near triple collision along the corresponding Lagrange road, at points L4 or L5. To account for these oscillations, take a positive integer N and call an eclipse sequence "N-long" if every time a number occurs in the sequence it occurs at least N times in a row. For example, the sequence 1112222333332222 is 3-long, but it is not 4-long, because there are only three 1s in a row.

Here, finally, is our main theorem: Consider the three-body problem with small nonzero angular momentum epsilon and masses within a large open range. Then there is a large positive integer N with the following significance. If we choose any eclipse sequence whatsoever— which is N-long—then there is a corresponding solution to our three-body problem having precisely this eclipse sequence. If that sequence is made to be periodic, then so is the solution realizing it.

What about my original question? There was no large N mentioned there. I had asked about every eclipse sequence. But I did not tell you my real question. What I really wanted to know was whether or not I could realize any "topological type" of periodic curve, not any eclipse sequence. I was using the eclipse sequence as a convenient shorthand or way of encoding topological type, which is to say as a way of encoding the winding pattern of the loop around the three binary collision holes. The eclipse sequence representation of the topological type of a closed curve has redundancies: Many different eclipse sequences encode the same topological type of curve. Consider, for example, the topological type "go once around the hole made by excluding the binary collision 23." The eclipse sequence 23 represents this topological type. But so do the eclipse sequences 2223, 222223, and 2333. Whenever we have two consecutive crossings of the arc 2, we can cancel them by straightening out the meanders, making the curve during that part of it stay in one hemisphere or the other without crossing the equator (Figure 15). Indeed, we can cancel any consecutive pair of the same number that occurs in an eclipse sequence without changing the topological type of closed curve represented by the sequence.

To use our main theorem to answer my real question, note that by deleting consecutive pairs I can ensure that the eclipse sequence that encodes a given topological type never has two consecutive numbers

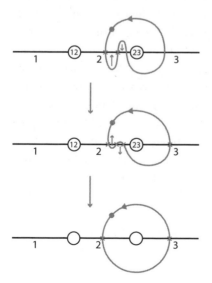

FIGURE 15. Eclipse sequence.

of the same type in it: no "11" or "22" or "33." Call such a sequence an admissible sequence. Now, take any admissible sequence, for example, 123232. Allow me to use exponential notation in writing down eclipse sequences, so, for example, $1^3 = 111$. Choose an odd integer n at least as big as the number N of our main theorem. Replace the admissible sequence by the longer sequence $1^n 2^n 3^n 2^n 3^n 2^n$ and continue it periodically. This longer sequence represents the same originally chosen topological type because n is odd. Our theorem says that this longer sequence is realized by a periodic solution. This periodic solution represents our original topological type 123232.

What's Next?

We still have much left to do. When I originally posed my question almost 20 years ago, I only wanted solutions having zero angular momentum. But evidence is mounting that the answer to my question in the case of zero angular momentum is "no." We have some evidence that even the simplest nonempty periodic sequence 23 is never realized by a periodic solution to the equal-mass, zero angular momentum three-body problem.

Our main question as posed here, even for angular momentum epsilon, remains open because our theorem allowed us to realize only sequences that are N-long for some large N. We have no clue, for example, how to realize admissible sequences, that is, sequences with no consecutive numbers of the same type.

At the end of the day, we may be no closer to "solving" the three-body problem in the traditional sense, but we have learned quite a lot. And we will keep at it—this problem will continue bearing fruit for those of us who are drawn to it. It turns out that new insights are still possible from one of the classic quandaries in mathematical history.

More to Explore

A Remarkable Periodic Solution of the Three-Body Problem in the Case of Equal Masses. Alain Chenciner and Richard Montgomery in *Annals of Mathematics*, 152(3), 881–901, November 2000.

Realizing All Reduced Syzygy Sequences in the Planar Three-Body Problem. Richard Moeckel and Richard Montgomery in *Nonlinearity*, 28(6), 1919–1935, June 2015.

The Intrigues and Delights of Kleinian and Quasi-Fuchsian Limit Sets

CHRIS KING

My research in dynamical systems has led me inexorably from the better-known fractal dynamics of Julia and Mandelbrot sets, including those of the Riemann zeta function [3], to the more elusive forms of Kleinian limit sets.

Julia sets and their universal atlases—Mandelbrot sets—have become ubiquitous features of the mathematical imagination, demonstrating the power of computer algorithms to generate, in scintillating detail, visualizations of complex dynamical systems, complementing theory with vivid counterexamples. Gaston Julia's original explorations, however, took place abstractly, long before the explosion of computing technology, so it was impossible to appreciate their fractal forms, and it was only after Benoit Mandelbrot popularized the set that now bears his name that the significance of Julia's work again became recognized. It is a historical irony, pertinent to this article, that the Mandelbrot set was actually discovered by Robert Brooks and Peter Matelski [1] as part of a study of Kleinian groups. It was only because Mandelbrot was a fellow at IBM that he had the computing resources to depict the set, in high resolution, as a universal multifractal atlas.

Julia sets are generated by iterating a complex polynomial or a rational or other analytic function, such as an exponential or harmonic function. For a nonlinear function, exemplified by the quadratic iteration $z \rightarrow f_c(z) = z^2 + c$, successive iterations either follow an ordered pattern, tending to a point or periodic attractor, or else in a complementary set of cases behave chaotically, displaying the butterfly effect and other features of classical chaos. The Julia set is the set of complex values on which the iteration is chaotic, and the complementary set, on which it is ordered, is named after the Julia set's codiscoverer, Pierre

Fatou. The Mandelbrot set M, Figure 1(h), becomes an atlas of all the Julia sets J_c as a result of beginning from the critical point $z_0 : f'(z_0) = 0$ (the last point to escape to infinity) and applying the iteration for every complex c-value. In a fascinating demonstration of algorithmic topology, if $c \in M$, then the fractal Julia set J_c of c is topologically connected. Otherwise, J_c forms a totally disconnected fractal dust, or Cantor set, with the most ornate and challenging examples lying close to, or on, the boundary of M.

Alongside images of complex Julia and Mandelbrot fractals that can be found on the Internet, you will also find a more esoteric class of fractal sets that often look like medieval arboreal tapestries, and go variously by the names Kleinian and quasi-Fuchsian limit sets, but the programs that generate them are much more difficult to find—to such an extent that I resolved to generate accessible multiplatform versions in the public domain to enable anyone who so desires to explore them.

To make these fascinating systems freely accessible and to aid further research, I have therefore developed an interactive website, *Kleinian and Quasi-Fuchsian Limit Sets: An Open Source Toolbox* [4], with freely available cross-platform software, including a MATLAB toolbox and a generic C script, one hundred times faster, that compiles and generates images on any GCC-compatible operating system, as well as a MacOS viewing application, enabling full dynamical exploration of the limit sets.

These limit sets also have an intriguing history, which I sketch only briefly, since it is elucidated in definitive and engaging detail in *Indra's Pearls*. Interspersed below are references to specific pages in *Indra's Pearls* at which the reader can find an expanded discussion of each topic.

In this field, we again witness an evolution whereby research and discovery have at first been driven by abstract theoretical advances, which are then followed by the emergence of innovative computational approaches that yield critical examples displaying the richness and variety that give the theory its full validation.

The field of Kleinian groups was founded by Felix Klein and Henri Poincaré; the special case of Schottky groups had been elucidated a few years earlier by Friedrich Schottky. The ensuing story, as depicted in *Indra's Pearls*, begins with a visit to Harvard by Benoit Mandelbrot that leads to David Mumford setting up a computational laboratory to explore "some suggestive 19th century figures of reflected circles which

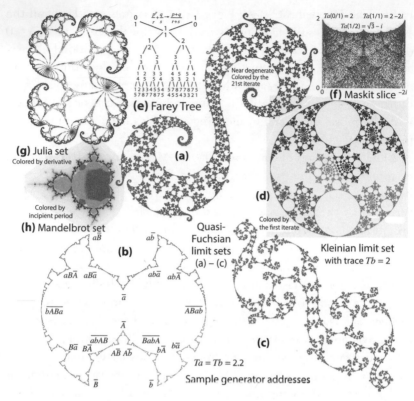

FIGURE 1. Julia sets and Kleinian limit sets (a)–(c) are two classes of complex fractals, the former the set on which a nonlinear function is chaotic, and the latter the limit set of two interacting Möbius transformations. Just as Julia sets (g) have the Mandelbrot set (h) as atlas, limit sets with $T_b = 2$, as in Figure 3, have the Maskit slice (f) as atlas, with the critical states on the upper boundary. See also color insert.

had fascinated Felix Klein." Caroline Series, along with David Wright, assisted by Curt Mullen, who had held summer positions at IBM, and other collaborators such as Linda Keen, then began to investigate computational exploration of these limit sets. In David Wright's words:

Take two very simple transformations of the plane and apply all possible combinations of these transformations to a point in the plane. What does the resulting collection of points look like?

Indra's Pearls continues this human and mathematical journey in elegant detail, noting the contributions of many researchers, including Bernard Maskit and Troels Jørgensen, to the unfolding exploration of the area.

A Mathematical Nexus

In contrast to Julia sets, Kleinian limit sets are algebraic "attractors" generated by the interaction of two Möbius maps a, b and their inverses $A = a^{-1}$, $B = b^{-1}$. Möbius maps [4, p. 62] are fractional linear transformations operating on the Riemann sphere $\hat{\mathbb{C}}$ represented by complex matrices; see Figure 2(a). Composition of maps is thus equivalent to matrix multiplication:

$$z \mapsto a(z) = \frac{pz + q}{rz + s} \Leftrightarrow \begin{bmatrix} p & q \\ r & s \end{bmatrix} \begin{bmatrix} z \\ 1 \end{bmatrix} : ps - qr = 1 \tag{1}$$

Möbius transformations can be elliptic, parabolic, hyperbolic, or loxodromic,[1] in terms of their trace $T_a = p + s$, as complex generalizations of the translations, rotations, and scalings of affine maps of the Euclidean plane. They map circles to circles on $\hat{\mathbb{C}}$, counting lines in the complex plane as circles on $\hat{\mathbb{C}}$. Möbius maps are thus simpler in nature than analytic functions and are linear and invertible, so a single Möbius map does not generate an interesting fractal. However, when two or more Möbius maps interact, the iterations of points in $\hat{\mathbb{C}}$, under the two transformations and their inverses, tend asymptotically to a set that is conserved under the transformations, often a complex fractal, called the *limit set* of the group. When the two transformations acting together form certain classes of groups, including Kleinian, quasi-Fuchsian, and Schottky groups, their limit sets adopt forms with intriguing mathematical properties.

Because they form an interface between widely different mathematical areas, their dynamics draw together diverse fields, from group theory through topology, Riemann surfaces, and hyperbolic geometry, to chaos and fractal analysis, perhaps more so than any related area. Depending on the particular group, its action[2] may transform the topology of the Riemann sphere into that of another kind of topological surface, such as a two-handled torus, sometimes with pinched handles; see Figure 2(e)–(g).

A Kleinian group is a discrete subgroup[3] of 2×2 complex matrices of determinant 1 modulo its center.

A Schottky group [4, p. 96] is a special form of Kleinian group, consisting of two (or more) Möbius maps a and b on $\hat{\mathbb{C}}$, each of which maps the inside of one circle, C, onto the outside of the other, D. If the two pairs of circles are entirely disjoint, as in Figure 2(b), the group action induces a topological transformation of $\hat{\mathbb{C}}$, generating a multihandled torus as in Figure 2(e).

The group interactions, when discrete, induce tilings in the complex plane, which also illustrate transformations of hyperbolic geometry.[4]

As the parameters of these maps vary, the limit sets approach a boundary in parameter space, where the limit set can become a rational circle packing, as in Figure 3(a)–(c), or irrationally degenerate, Figure 3(f), (g), forming a space-filing fractal, beyond which the process descends into chaotic dynamics, Figure 6(d), as the groups become nondiscrete and their orbits become entangled. Like the Mandelbrot set, for certain classes of limit set, one can generate an atlas, called the Maskit slice, Figure 1(f), whose upper boundary contains these limiting cases.

An insightful way to portray their actions is to determine what the Möbius transformations do to key circles involved in the definitions of a and b. A good starting point is to consider pairs of Schottky maps. If all the circles are disjoint when they are transformed by both maps, then their images form a cascade of circles, which tend in the limit to a Cantor set; see Figure 2(b). As the parameters p, q, r, and s of each are varied, so that circle pairs come together and touch, as in Figure 2(c), the Cantor set transforms into a circular limit set.

There is an alternative way, independent of Schottky groups, to uniquely define the transformation matrices via their traces T_a, T_b, where $T_a = p + s$, as in (1), if we focus on maps for which the commutator $aba^{-1}b^{-1}$ is parabolic, with $T_{aba^{-1}b^{-1}} = -2$, inducing a cancellation in equation (2) below. Since traces are preserved under conjugacy of mappings, $b = cac^{-1}$, we can simplify the discussion by restricting our attention to conjugacy classes, reducing the number of free variables in the two matrices to a unique solution. This process, affectionately known in *Indra's Pearls* as "Grandma's recipe" [4, p. 227], enables us to explore new classes of limit sets, as illustrated in Figure 2(d) for the Apollonian gasket, which can also be represented via four circles, all of which touch, as shown near the center.

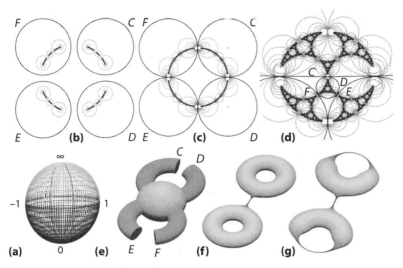

FIGURE 2. (a) The Riemann sphere, i.e., the complex plane closed by a single point at infinity. (b) Schottky maps forming Cantor dust. (c) Tangent circles forming a circular limit set. (d) Four circles, all tangent, forming an Apollonian gasket. (e) Schottky maps perform a surgery of $\hat{\mathbb{C}}$ to form a double-handled torus. (f) If the commutator is parabolic, as in quasi-Fuchsian limit sets, which form a fractal topological circle, or a Jordan curve, whose complements quotient by the group into a pair of Riemann surfaces, the double torus becomes pinched. (g) When one or both of a and b, or their words, are also parabolic, as in the gasket (d) and those in Figure 3, one or two further pinches occur. See also color insert.

The asymmetric nature of the two matrix mappings generated by Grandma's recipe can be seen in Figure 1(b), (d), where both are colored by the first iterate. Although the commutator is parabolic (resulting in the single pinch we see in Figure 2(f)), the generators a and b can be loxodromic and can tend to be parabolic as well, with the transition from the quasi-Fuchsian set of Figure 1(b), where $T_a = T_b = 2.2$, to the Apollonian gasket of Figure 3(a), where $T_a = T_b = 2$, resulting in the two further pinches in Figure 2(g) and in each of the examples in Figure 3, where b is parabolic with $T_b = 2$.

Because the limit set is the place where all points are asymptotically mapped under the interaction of the transformations and their inverses, one can simply pick each point in the plane and repeatedly map it by a random sequence of the two transformations and their inverses, and these points will become asymptotically drawn to the limit set. The trouble with this process is that although it can crudely portray any limit set, including chaotic nondiscrete examples, the asymptotic iterates are distributed exponentially unevenly, so that some parts of the limit set are virtually never visited and the limit set is incompletely and only very approximately portrayed, as in Figure 6(d).

In contrast to the nonlinear functions of Julia sets, where depiction algorithms, such as the modified inverse iteration of Figure 1(g) and distance estimation, depend on analytic features such as derivatives and potential functions, an accurate description of Kleinian group limit sets requires taking full strategic advantage of the underlying algebraic properties of the group transformations.

Descending the Spiral Labyrinth

The depth-first search (DFS) algorithm to depict the limit sets is ingenious and extremely elegant [4, p. 141]. The aim is to traverse the algebraic space of all word combinations of the generators a, b, a^{-1}, b^{-1} in a way that draws a continuous piecewise linear approximation to the fractal at any desired resolution.

To do this, we generate a corkscrew maze in which we traverse deeper and deeper layers of the search tree, turning right at each descent, and then enter each tunnel successively, moving counterclockwise around the generators and retreating when we reach the inverse of the map by which we entered, after three left turns, since the map and its inverse cancel. At each node, we retain a record—our Ariadne's thread—of our journey down, by multiplying the successive matrices as we descend, and we make a critical test: The parabolic commutator has a single fixed point, which represents an infinite limit of cyclic repeats of the generators. If we apply the composed orbit matrix to the fixed point of the clockwise commutator and do the same for the counterclockwise commutator's fixed point and these two are within an epsilon threshold, this means that going the "opposite way" around the local fractal leaves us within resolution, so we draw a line between the two and terminate the descent, retreating and turning into vacant tunnels and exploring them, until we find ourselves back at the root of the tree.

Using the fixed points has the effect of producing a theoretically infinite orbit of transformations that can carry us to any part of the limit set, and because we are traversing generator space systematically, we will traverse parts of the limit set that are visited exponentially rarely in a random process.

Navigating the Maskit Slice

If we focus on limit sets in which b is parabolic with $T_b = 2$ and only T_a varies, this gives us a complex parameter plane called the Maskit slice [4, p. 287], Figure 1(f), forming an atlas of limit sets, just as the Mandelbrot set does for Julia sets, so we can explore and classify their properties. To define locations on the slice, we need to be able to determine the traces T_a that correspond to rational cases involving fractional motion. Key to this are the generator words formed by the maps. For example, the set in Figure 3(c) labeled "3/10" has a generator word $a^3Ba^4Ba^3B$ symbolically expressing the 3/10 ratio (three B's to ten a's), which is also parabolic. For general fractional values p/r we can determine the traces recursively from two key relations: extended Markov and Grandfather [4, p. 192]:

$$T_{aba^{-1}b^{-1}} = T_a^2 + T_R^2 + T_{aR}^2 - T_a \cdot T_R \cdot T_{aR} - 2 \tag{2}$$

$$T_{mn} = T_m \cdot T_n - T_{m^{-1}n} \tag{3}$$

We can use these relations to successively move down the Farey tree of fractions [4, p. 291], see Figure 1(e), because neighbors on the tree have generator words combining via Farey mediants $w_{p+q/r+s} = w_{p/r}w_{q/s}$, and so the Grandfather identity implies $Tw_{p+q/r+s} = Tw_{p/r}Tw_{q/s} - Tw_{p-q/r-s}$, because in the last term, the combined word $m^{-1}n$ has neatly canceling generators in the middle. Since $T_{aba^{-1}b^{-1}} = -2$ in (2), for p/r, this reduces to solving an rth-degree polynomial.

If we plot all the polynomial solutions to the p/r limit sets for $r \leq n$, we have the Maskit slice [4, p. 287], Figure 1(f), whose upper boundary contains each of the p/r trace values as fractal cusps. Above and on the boundary are discrete groups, which generate fractal tilings of the complement in $\hat{\mathbb{C}}$, while values within the slice produce nonfree groups or chaotic nondiscrete sets, which can generally be portrayed stochastically, Figure 6(d), but not by depth-first search.

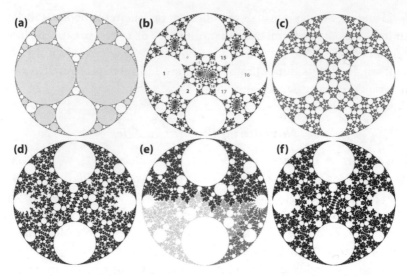

FIGURE 3. (a) Apollonian gasket fraction 0/1, $T_a = 2$, shaded to highlight separate components in the mapping. (b) The 1/15 limit set with the 15 steps of the parabolic word $a^{15}B$ (numbers 1–15) in red. Intriguingly, as $1/n \to 1/\infty = 0$, the limit sets do not tend to the gasket, although their traces and matrices do, since the limit set has an additional fourfold symmetry of rotation by $\pi/2$ already evident in (b). (c) The 3/10 set colored by the last iterate. (d) A free discrete group. (e) Singly degenerate $(LR)^\infty$ Farey set of the golden mean. (f) Spirally degenerate $(L_{10}R)^\infty$ set. See also color insert.

The rational cusps [4, p. 273] provide intriguing examples of circle packings with deep links to hyperbolic geometry. However, irrationally degenerate space-filling limit sets remained enigmatic until a key example trace value was found [4, p. 314], named after Troels Jørgensen, who posited the existence of such limit sets [2]. The solution is completely natural, the golden mean limit of the Fibonacci fractions arising from a repeated LR move on the Farey tree (e.g., $1/1 - L \to 1/2 - R \to 2/3$ in Figure 1(e)), resulting in the limit set of Figure 3(e). We can approach such values using Newton's method to solve the equations of the ascending fractions.

This method leads to further cases, such as spiral degeneracy [4, p. 320], where we repeat a ten-to-one set of moves $L_{10}R$ (ten left turns followed by one right turn) up the left-hand slope of the Maskit slice upper boundary and into a spiraling sequence of ever smaller cusps in the slice, giving rise to the limit set of Figure 3(f). Since the set of irrational numbers is uncountable, while the complementary set of rational numbers is countable, such limit sets represent the overwhelming majority of cases, although they are harder to access.

Double Degeneracy at the Edge of Chaos

The above examples are singly degenerate because only half the region is engulfed, and the question arises whether it is possible to find a pair of traces generating a limit set that is entirely space-filling, permeating the complex plane with a fractal dimension of 2. This quest became a mathematical epiphany [4, p. 331]. It is possible to generate a limit set conjugate to Troels's example in the following way. To make an *LR* journey down the Farey tree to a higher Fibonacci fraction, such as 987/1597, about the middle, we find 21/34 and 13/21. It turns out that the words in terms of a and B for these steps also form generators that can be used in reverse to define a and B. If we apply Grandma's algorithm to their traces calculated by matrix multiplication of the a and B words, we get another limit set conjugate to the original, as shown in Figure 4(a).

The reasoning that these two midpoint traces are very similar led to the idea that a doubly degenerate limit set corresponding to the golden mean might arise from simply applying endless *LR* Farey moves

$$(a, B) - L \to (a, aB) - R \to (a^2B, aB)$$

that leave the generators unchanged, $(a, B) = (a^2B, aB)$, giving two equations that combined with Grandfather and Markov lead to the conjugate trace solutions $(3 \pm 3^{1/2}i)/2$, as shown in Figure 4(b).

An intriguing feature is that in addition to a, b, there is a third induced parabolic symmetry c, because the $(LR)\infty$ move must arise from a Möbius map conjugating the limit set to itself, which has the same fixed point as the parabolic commutator and conjugates with it, although it performs a distinct "translation," flipping the sets bounding the jagged line connecting the centers. Robert Riley [6] has shown that the three transformations result in a gluing of 3D hyperbolic space, via the

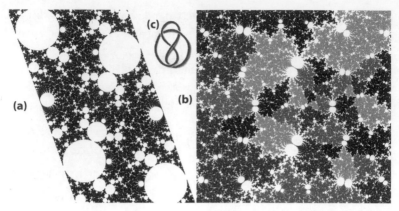

FIGURE 4. (a) Limit set conjugate to the golden mean limit set. (b) The doubly degenerate space-filling limit set of fractal dimension 2, colored by the third iterate to highlight the fractal line connecting centers. (c) The figure-eight knot, whose 3D complement is glued from hyperbolic 3-space by (b). See also color insert.

group's discrete 3D "tiling," to become the 3-sphere with the figure-eight knot (Figure 4(c)) removed [4, p. 388].

To unravel the spirally degenerate case, I examined the recurrence relations from the endless $(L_{10}R)^\infty$ move and found that they could be exploited to produce a pair of traces by solving a system of 11 equations:

$$T_a = 1.936872603726 + 0.021074052017i$$

$$T_b = 1.878060670271 - 1.955723310188i$$

When these are input into the DFS algorithm, we have the double spiral degeneracy shown in Figure 5.

The Bondage of Relationships

There is a further enchanting collection of limit sets [4, p. 353] for which the group is nonfree and has a commutator that squares to the identity: $(aba^{-1}b^{-1})^2 = I$. This situation causes the trace to be 0, but an extended Grandma's algorithm [4, p. 261] applies, the recurrence relations (2) and (3) can be solved to give a polynomial, and the limit sets

FIGURE 5. The "Holy Grail": double spiral degeneracy, with the traces reversed in the second image. The convergence to space-filling is very slow, and the full resolution image took 29 hours in generic C and would have taken 112 days in interpreted MATLAB code. Although the traces are no longer conjugates, the sets are clearly degenerate on both complement components in the same way as depicted in Figure 4.

can be depicted by DFS, as long as generator words that short-circuit to the identity are treated as dead ends.

This notion requires an automatic group, involving a lookup table (automaton) in which all the growing word strings that could end in a short circuit as we descend the search tree are accounted for by iteratively composing their states and exiting when a death state is reached. Figure 6(a)–(c) shows three such limit sets, with the "inner" regions highlighted in yellow to distinguish them from the complementary white regions, from which they are isolated by the limit set.

Falling into the Chaotic Abyss

One can also freely explore the wilder limit sets using the stochastic algorithm, which recursively maps chosen points, such as Möbius transformation fixed points, using a random sequence of the four generators. This exercise can aid in visualizing limit sets in lower fidelity without any restriction, including nonfree and nondiscrete chaotic systems, as illustrated in Figure 6(d).

Chris King

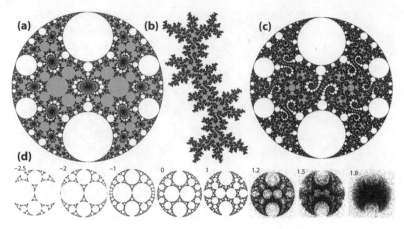

FIGURE 6. Nonfree limit sets: (a) $T_a(1/60)$, $T_b = 2$. (b) A quasi-Fuchsian "dragon." (c) $L_{10}R$ spirally degenerate. (d) A succession of nonfree and non-discrete chaotic limit sets for which $T_a = T_b = 2$ under a varying commutator trace, using the stochastic algorithm. See also color insert.

Notes

1. Parabolic maps have $T_a^2 = 4$ in the complex plane C and act like translations, with one fixed point on R; elliptic maps have $0 < T_a^2 < 4$ and behave like rotations on R; hyperbolic and loxodromic maps have real and complex values lying outside these values on C and behave like (spirally) expanding and contracting versions of scalings on R.

2. This new topology arises from the quotient of the complement of the limit set under the action of the group.

3. A discrete group is a group of transformations of an underlying space possessing the discrete topology. Discrete groups thus appear naturally as the symmetries of discrete structures, such as tilings of a space.

4. The group of all Möbius transformations is isomorphic to the orientation-preserving isometries of hyperbolic 3-space $H3$, in the sense that every orientation-preserving isometry of $H3$ gives rise to a Möbius transformation on Ĉ and conversely. Hence, group actions of such transformations can generate hyperbolic tilings. Maurits Escher's "Angels and Demons" tiling is a representation of the modular group, which is also manifest in the symmetries of the gray Apollonian gasket disks [4, p. 206] in Figure 3(a), and it has a similar topological pinching [4, p. 216] to that shown in Figure 2(g).

References

[1] Robert Brooks and Peter Matelski. The dynamics of 2-generator subgroups of PSL(2,C). In Irwin Kra, ed., *Riemann Surfaces and Related Topics: Proceedings of the 1978 Stony Brook Conference.* Princeton University Press, Princeton, NJ, 1981.

[2] Troels Jørgensen. Compact 3-manifolds of constant negative curvature fibering over the circle. *Ann. Math. Ser 2* 106(1) (1977), 61–72.

[3] Chris King. Fractal geography of the Riemann zeta function, arXiv:1103.5274, 2011.

[4] Chris King. Kleinian and quasi-Fuchsian limit sets: An open source toolbox. Available at http://dhushara.com/quasi/, 2018.

[5] David Mumford, Caroline Series, and David Wright. *Indra's Pearls: The Vision of Felix Klein.* Cambridge University Press, 2002.

[6] Robert Riley. A personal account of the discovery of hyperbolic structures on some knot complements. *Expo. Math.* 31 (2013), 104–115.

Mathematical Treasures from Sid Sackson

JIM HENLE

*Games mean many things to many people; to me they
are an art form of great potential beauty.*
—*Sid Sackson,* Beyond Tic Tac Toe

Sid Sackson (1920–2002) was a professional game inventor. He was
a seminal figure in twentieth-century game design, with pioneering
board games, card games, party games, paper-and-pencil games, solitaire games, and word games. In this column I will argue:

1. That (many) games are mathematical structures.
2. That games and mathematics share many of their most important aesthetics.
3. That Sid Sackson created games that pose interesting mathematical problems.

But before I do, let me tell you a little about Sid. Sid Sackson was in-
venting games as a child. He designed games all his life, though he had
a day job as an engineer into his 50s.

Acquire was probably his most successful game. It's a business game
with hotels, stocks, and mergers, with geometric elements borrowed
from war games. *Acquire* launched 3M's collection of bookshelf games.

Many of Sackson's games, *Twixt* and *Sleuth*, for example, were published
by 3M. More than a few were published in Germany by Ravensburger,
Piatnik, and other companies. One of these German games, *Focus*, won
the Spiel des Jahres (game of the year) and Essen Feather awards.

Sackson lived in the Bronx. Game designer Bruce Whitehill said
of him:

Math was always his passion; he wanted to be a math teacher
when he was in college, but was told his voice wasn't good for

teaching, so he went into engineering. He loved to dance . . . He hobnobbed with celebrities sometimes (and loved it), such as Tony Randall. Shari Lewis came to the house once. Sid shared a TV spotlight with Omar Sharif.

I was struck by another of Whitehill's recollections. After Sackson had passed away, he asked Bernice, Sid's wife of 61 years,
"Are there any funny stories?"
She replied, "I wish there were."[1]

Mathematical Structures

For centuries philosophers have disputed the definition of mathematics. We don't have to touch that argument. We have only to agree what a mathematical structure is. I have a simple definition, one that is understandable to nonmathematicians and at the same time (I hope) is amenable to philosophers of mathematics:

A **mathematical structure** is anything that can be described completely and unambiguously.

Whatever your definition of mathematics, the field definitely involves logical deduction. This is the case even in such fields as probability and quantum logic. Mathematical structures as defined above are both sufficient and necessary as subjects for logical deduction.

Most games are mathematical structures. The rules of a game, if properly written and useful, describe the game completely and unambiguously. I don't mean to claim here, as formalist philosophers do, that mathematics is in its entirety simply a game.[2] Clearly much mathematics came into being to help understand and capture aspects of the physical world. But then there are games with the same purpose—war games, social games, economic games, etc.

This article, however, is about those mathematical structures that are intended to please, and most games are so intended.

Game Aesthetics

If we restrict our attention to games of strategy and tactics, the characteristics that make a game especially intriguing are exactly those that attract us to a mathematical theory. In mathematics, we appreciate a deep result with minimal hypotheses. Quite similarly, a game is especially attractive if its play is complex and difficult to master, yet the rules are simple and easy to follow.

Tic-tac-toe has simple rules, but the strategy is not challenging and is easily mastered. Whist is a complex game, but the rules are not simple; they are complicated and fussy.

In contrast, the Japanese game of Go is an exemplar of game beauty. The rules are few and elegant (much simpler than those of chess). Yet the strategy is enormously complex. As with chess, people devote their lives to improving their play of Go.

The game of soccer—known internationally as "football"—is often called the "beautiful game." While not a mathematical game, I believe this description reflects the same aesthetics.

Sackson's "Cutting Corners"

I first met the work of Sid Sackson through his book *A Gamut of Games*.[3] It contains 38 games, a few old, many by friends, and 22 by Sackson himself. They are all gems. Whenever I see this book at a used bookstore or at a yard sale, I buy it. Martin Gardner said that *A Gamut of Games* was "the most important book on games to appear in decades."

Cutting Corners is a Sackson game from *Gamut*. It illustrates the simplicity and complexity of his work. It is played on a small square edged in two colors. Lacking color, we will use solid lines and dashed lines.

The dashed line is the "color" of Player I, the player who goes first. The solid line represents Player II.[4] A move consists of drawing an ell-shaped line that starts perpendicularly from an edge, makes a right-angled turn, and ends on another edge. Each move must either cross a line of the other sort or end on an edge of the other sort.

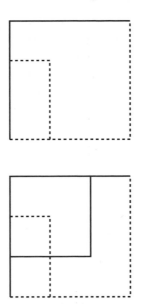

Additionally, the *n*th move, for every *n*, must cross exactly $n - 1$ lines.

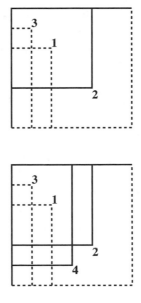

The game ends after two more moves (a total of six).

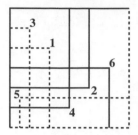

To determine the winner, examine each region created by the lines. Mark it for Player I if it has more dashed edges than solid edges. Similarly, mark it for Player II if it has more solid edges than dashed.

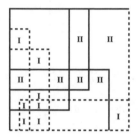

The winner is the player who has more marked regions. In this case, Player I (dashed lines) wins.

Complex! But we can trim the game down by reducing the number of moves.

It's pretty easy to see that Player II has a winning strategy in the two-move game. In the three-move game, Player I, who gets to draw two lines, can always beat Player II, who only draws one. This is what you would expect, but it still takes checking. There is only one possible first move, but there are three possible second moves.

The game to tackle next, if you want a challenge, is the four-move game, where each player makes two moves. My money is on Player II, but the game doesn't look easy to analyze. I counted 23 three-move patterns. You might have to explore all of them.

By the way, is it possible to have a tie in this game? I don't know the answer. Sackson doesn't mention the possibility. It seems like an intriguing mathematical question.

Sackson's "Hold That Line"

Hold That Line is played on a 4 × 4 array of dots:

A play consists of drawing a straight line from one dot to another:

Each line after the first must begin at a free end of a previously drawn line:

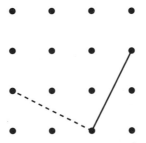

Lines may not cross. The last player to make a legal move is the *loser*.

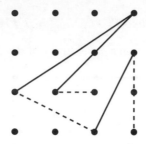

It's a tricky game and clearly a subject for logical analysis. To start, let's examine the rules as applied to smaller boards.

It's easy to see that in the game with a 2 × 2 array, Player I has a winning strategy—just draw the diagonal. It's more work, but it's not hard to show that Player I has a strategy in the 3 × 2 game. This initial move works:

A similar move also works for Player I in the 4 × 2 game. It doesn't work, though, for the 5 × 2 game. A different strategy works, however. Can the reader find one?

Hold That Line is a *misère* game; that is, the last player to move loses. Why did Sackson choose to make the last player the loser? Nothing is written about this, but I'm sure it's because there is a simple strategy for the game where the last player to move wins. Player I starts with the diagonal,

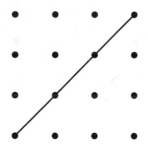

and then mimics Player II's moves symmetrically, for example:

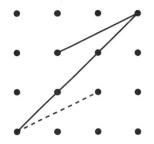

If Player II has a legal move, then Player I will have one too, all the way to the end. Thus, Player I will make the last move.

Is the full (4 × 4, misère) Hold That Line too difficult to analyze? All I can say is that it has so far defeated me. If you see a way, let me know! I will report any discoveries at www.math.smith.edu/~jhenle /pleasingmath/. *Note added for this anthology*: Jindřich Michalik of Charles University has just proved that the player going first has a winning strategy for the 4 × 4 game of Hold That Line.

"Springerle"

Sackson wrote a series of inspired paperbacks, each containing a collection of his games with a common theme. One book featured cooperative games; another featured solitaire games; another featured calculator games; yet another featured word games.

The two-person game Springer is from one of these books, *Beyond Tic Tac Toe*.[5] The name of the game is a reference to the German painter and printmaker Ferdinand Springer, since some of his art resembles the board of Springer at the end of a game. In German, *Springer* is the word for a chess knight, which serendipitously is the logo of Springer, publisher of *The Mathematical Intelligencer*. This Ferdinand Springer, it should be noted, is no relation to Julius Springer, the founder of Springer Publishing, or his son Ferdinand.

Springer has a large board and a somewhat elaborate set of rules for what constitutes winning. I have pared this down to a simpler game, which I call Springerle.

Springerle is played on this small board:

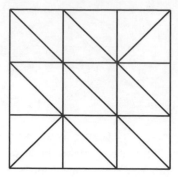

Players take turns coloring triangles with their color (here speckled for Player I, solid for Player II), subject to the restriction that no two of a player's triangles can share a side.

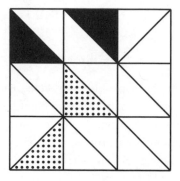

A player wins if his or her triangles form a network connected by vertices that contains part of each side of the square. Here is a win for Player II (solid).

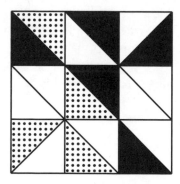

One would think that the player going first has an advantage, but the restriction about neighboring triangles makes a difference. For example, if Player I starts out here,

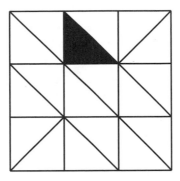

then Player II can make sure Player I loses just by playing here,

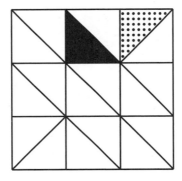

because now it's impossible for Player I to reach the top edge.

The example above also shows that a draw is possible, that is, neither player wins. I suspect, however, that Player I can force a win. I don't have a proof, though. Can someone help?

One could call Springerle a "crossing game," like Hex.[6] There is a wonderful "strategy-stealing" proof that Player I has a winning strategy for Hex. It works by showing that if Player II had a winning strategy, Player I could "steal" it, that is, use it successfully to win a game of Hex. The above example shows that wouldn't work here.

The fun of Springerle lies in the ease with which a player can be blocked from touching an edge. That makes it difficult to design an appropriate board of a large size. This might explain Sid Sackson's more elaborate rules for his 9 × 9 board.

"Bowling Solitaire"

This is a solitaire game using only 20 cards. Take two each of 2 through 10 and two aces—the suits don't matter; it's only the denomination that counts—and think of the ace as representing the number 1. I will assume that the reader is familiar with the way bowling is scored.

Shuffle the cards and deal them out in the pattern of bowling pins and form piles of five cards, three cards, and two cards face down, each with the top card turned up.

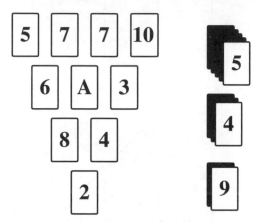

The object, of course, is to knock down pins. You may knock down one, two, or three adjacent cards using a card from one of the piles if the sum of the pin cards has the same *units* digit as the card from the pile. For example, you can use the 4 card in the second pile to remove cards 6 and 8 (6 + 8 = 14).

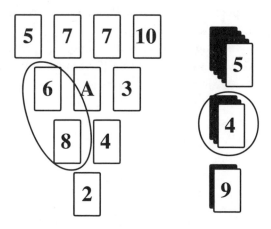

The **4** could instead be used to remove the ace and 3, or the 4, 3, and 7. But it can do only one of these. When knocking down pins, the pin cards and the card from the pile are discarded, and a new card is turned up on the pile.

Special rule: You aren't permitted to knock down any pins in the back row on the first card.

You continue to knock down pins until you are no longer able—or no longer wish to. That was your first ball (this is bowling, remember)? At this point, you discard the top cards of the piles and turn over the new top cards.

Continue knocking down pins; this is your second ball. When you stop, discard the top cards of the piles and turn over the new top cards. Knock down pins once more. That's your third ball.

Let's think about the setup we have dealt here. We could knock out the 6 and 8 with the 4. Then we could knock out the 2, 4, and 3 with the 9. Let's do it. First the 6 and the 8.

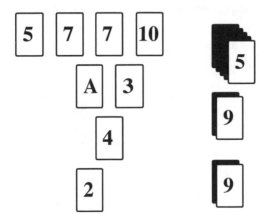

Ooh! Under the 4 was a 9. Now let's use one of the 9s.[7]

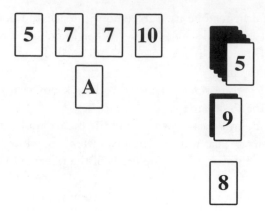

We could knock down the ace and two 7s with the 5. But then we would never get rid of the remaining 5, because all the 5s would be gone. We could instead just knock down the 5 with the 5 and then the 7 and the ace with the 8. But that would leave us with a 7 and a 10, which we could never knock down (there are no more 7s).

The last possibility is to use the other 9 to knock down the 5 and two 7s. That leaves us a chance of getting a strike or a spare. We *might* turn up a 10 and an ace. Here we go.

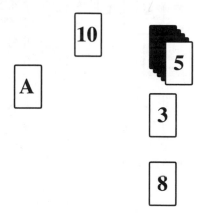

Phooey. That's the end of the first ball. We knocked down eight pins.

What do we get for the next ball?

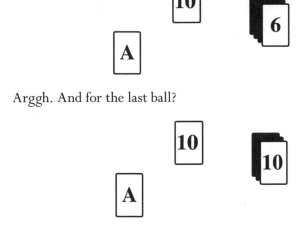

Arggh. And for the last ball?

Good! Okay. Knock down the 10 and turn the card. There are two cards underneath. One is a 2 and the other is an ace. We have a 50% chance of getting a score of 10 for the frame.

Yes!

"The Last Word"

I'll close with one more Sackson game. It's not mathematical; it's a word game. I'm doing this to be nice. The game is a treat.

The fact that it's a word game means that it doesn't have the aesthetic of simplicity, since its set of rules contains an entire English language dictionary. Nonetheless, it is an amazing game, elegant in the mathematical sense. It is a challenging, exciting word game that takes only a single sheet of paper and a pencil. Play it on a plane ride, and the hours will pass quickly away.

The Last Word is played on a 9 × 9 grid. Start by filling in the middle nine squares with any letters you like. I usually pick up a book, find a page, and copy the first letters I see. We're on a plane, though; what do I see? "Fasten seat belts." Okay.

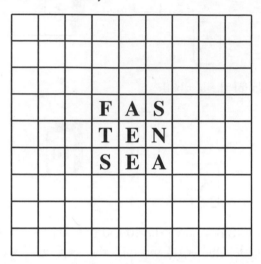

A move consists of placing a letter in a square next to a letter already placed. You get points by rearranging consecutive letters in a single direction to form a word. For example, by putting a P here,

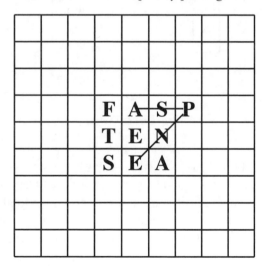

you can get the words SAP in one direction and PEN in another. You aren't allowed to jump over letters; that is, you couldn't use the S and the F without the A.

SAP and PEN are both three-letter words, but our score isn't the sum; it's the product. This move is worth nine points.

That's not the best first move, though. A T in the same place would get you FAST and TEN. That would be 12 points.

Still better would be an I here:

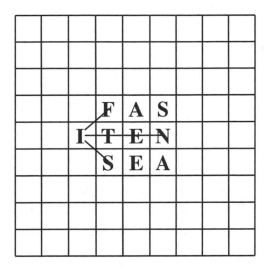

That's worth 16 points from IF, TINE, and IS.

What's especially nice about this game is that at later stages the scores get very high, hundreds of points, so that you can be losing by a large margin and catch up quickly. Do you see, by the way, that if Player I places the I there, the second player could score 24 points?

The game ends when all four sides of the square have been reached. As always, you can reach me at pleasingmath@gmail.com.

Notes

1. Everyone I've told this to agrees that this is a funny story.

2. But I might do that somewhere else. And maybe I did do that somewhere else. Ignore that.

3. Castle Books 1969, Pantheon 1982, and now available through Dover Publications.

4. I will use these identifications for all the games. They aren't always necessary, but they make the progress of the games easier to note.

5. Pantheon Books, 1975.

6. Invented independently by Piet Hein and John Nash.

7. This is not a carefully constructed example. I shuffled the cards and dealt them. This is *real*.

The Amazing Math inside the Rubik's Cube

Dave Linkletter

Next year marks 40 years of the Rubik's Cube first hitting toy shelves. Ever since its release, it has taunted almost a half billion tinkerers who think they can crack the confounding cube, only to be stymied by its maddening secrets. As we approach the Rubik's Cube's ruby anniversary (for real!) it is time to unpack the puzzle once and for all—with deep math. While the cube's literal insides may be made of plastic, its real guts are nothing but numbers. Let us dive in.

Breaking Apart the Blocks

Starting with some basics, a 3 × 3 × 3 Rubik's Cube has six faces, each a different color. The center of each face is attached to the core scaffold that holds the cube together, so they do not move, other than rotating in place. As a result, the same colors always end up opposite each other; on a standard cube, white is opposite yellow, red opposite orange, and blue opposite green.

Bust open a Rubik's Cube and you will see it is made of three types of building blocks. First, there is that central scaffold, connecting the center of each face. Then there are the cubies—the nickname for the little 1 × 1 × 1 blocks. The corner cubies have three colored sides, and the edge cubies have two. A Rubik's Cube has one core, eight corner cubies, and 12 edge cubies.

The immediate math to be done with those numbers is the total number of ways you can scramble a Rubik's cube: 43,252,003,274,489,856,000. Written in a more mathematical way, that number is $(3^8 8!)(2^{12} 12!)/12$. Here is how that comes together.

FIGURE 1. Rubik's Cube blocks (aka "cubies"). See also color insert.

The first term, 3^8, counts every way the eight-corner cubies can be rotated. A corner cubie can fit into its slot rotated three different ways. That is a factor of 3 for each of the eight corner cubies, so they multiply to 3^8.

Next is where each corner cubie goes. There are eight corner slots, so the first corner cubie has eight options. The second corner cubie is left with seven options, the next is left with six, and so on, down to the last corner cubie, which must go in the last corner slot. That yields the calculation $8 \times 7 \times 6 \times 5 \times 4 \times 3 \times 2 \times 1$, which is 8!, or "eight factorial."

Thus the first chunk, $(3^8 8!)$, counts every way the corner cubies can fit into the cube. The 3^8 is their orientations, while the 8! is their locations.

The next chunk, $(2^{12} 12!)$, is the same idea, now for the edges. Edges only have two orientations, so the 12 of them have a total of 2^{12} mixes of orientations. Then there are 12 locations, so 12! is the number of ways they can go to those spots.

What is left of the formula $(3^8 8!)(2^{12} 12!)/12$ is that division by 12. It relates to a fact about Rubik's Cube that is often felt, but not always understood. Here is a thought experiment (which perhaps you have done for real!) to illustrate:

Suppose you break open a Rubik's Cube, remove each cubie, and then put all the cubies back in random slots (with corner cubies only fitting in the corners, and edge cubies only in the edges). You get what looks like a normal scrambled cube, and so far we have counted every way you could do this, $(3^8 8!)(2^{12} 12!)$. Now, is it always possible to solve this jumbled cube, without breaking it apart?

The answer is no.

This is a trap that has caught many novice cubers. If you are practicing and you want to scramble a solved cube, you have to keep the cube intact and scramble it up manually. If you break it apart and reassemble the cubies randomly, there is actually only a 1 in 12 chance that it will be solvable.

The Answer Is in the Algorithms

Want to understand why that answer is 1 in 12? Well, there is a nice visual way to get a sense of it. A cube that has been broken and reassembled with its cubies randomly scrambled has equal chances of being solvable to one of the following representatives.

We have arranged these images to splay out the different factors leading to 12. Row 1 has normal corners. Rows 2 and 3 have one corner

FIGURE 2. Broken and reassembled cubes. The orange, yellow, and green sides (not shown) are solved as usual. See also color insert.

rotated in place. Column 1 has normal edges. Column 2 has one edge flipped in place. Column 3 has two edges swapped. Finally, column 4 has one edge flipped plus two edges swapped.

So the 12 cubes in Figure 2 cannot be transformed into one another. And there is no 13th arrangement that cannot transform into one of those 12. How do we know this?

There is a connection here with what can and cannot be done by moving the cube's faces. A sequence of moves is often referred to as an "algorithm" by cube enthusiasts. The sought-after algorithms are those that move just a few of the cubies while leaving the rest untouched. The limitations to the algorithms are the key to that number 12.

That 12 comes together from three factors being multiplied: $12 = 3 \times 2 \times 2$. We need to grapple with a factor of 3, and two factors of 2.

The factor of 3 comes down to this: There is an algorithm that twists each of two different corners, but there is *no* algorithm that twists a single corner (while leaving everything else unmoved). So if you grab a normal Rubik's Cube, pry out a single corner, and replace it twisted, it becomes impossible to solve, and you will have moved from the top left corner of our chart to one of the spots right below it.

However, if you repeat that process, and twist one more corner, it does not add a second factor of 3. Now that two corners are twisted, we can apply the algorithm that twists two corners, until at least one is fixed. If the other one happens to be fixed in the process, we got lucky, and now we are back to a solvable cube. Overall, the corners' orientations can go one of three ways.

The first factor of 2 is similar. There is an algorithm that flips, in place, each of two different edges, but no algorithm can flip a single edge in place. So any number of flipped edges can be shimmied down to a single edge, which winds up either flipped or not, for two possibilities.

The last factor of 2 actually involves edges and corners, though we showed it on the chart with edges. There is an algorithm that swaps two corners while also swapping two edges. There are no algorithms that can swap only a pair of corners, nor only a pair of edges.

If you have a cube, pry out two edges, and swap them, you jump by two columns on our chart—either between columns 1 and 3, or between 2 and 4. The same is true if you swap a pair of corners. But swapping a pair of edges *and* a pair of corners cancel each other out, since there is an algorithm to undo that.

With each factor in that division by 12 explained, you have the full picture on $(3^8 8!)(2^{12} 12!)/12$. There are $(3^8 8!)(2^{12} 12!)$ ways to put the cubies on the cube, but only one in 12 of those can be maneuvered to a solved cube. So $(3^8 8!)(2^{12} 12!)/12$ is the number of ways you can scramble a Rubik's cube, without breaking it apart.

The **Popular Mechanics** *Rubik's Cube Proof*

Now, if you are thinking inquisitively, you could desire proof for some of the claims in the last paragraphs. Is there some deeper math that can prove "there is no algorithm that flips one edge cubie in place without moving any other cubie"? You bet. Here is how that mathematical proof roughly goes:

> When a face of the cube is turned, four edge cubies are moved. Consider, for instance, an algorithm of 10 moves. For each cubie, follow it through the algorithm, and count how many times it is moved, and call that its cubie-moves count. Add up those numbers for every edge cubie, and the total must come to 40 cubie-moves, since each of the 10 moves adds four to the total.
>
> In general, any algorithm's total number of cubie-moves for the edge cubies must be a multiple of 4. Now for a critical pair of facts: If an edge cubie is moved an even number of times and returned back to the same slot, it will have the same orientation. Conversely, if an edge cubie is moved an odd number of times and put back in the same slot, it will be flipped.
>
> And yes, that pair of facts can be proven with *even deeper* math, but we will stop zooming in from here, in the name of this article eventually ending. You can also check the two facts experimentally and get a feel for why they are true. (For this proof, a 180-degree turn counts as two moves of each cubie involved.)
>
> Finally, consider a hypothetical algorithm that accomplishes the goal here, flipping one edge cubie in place without changing any other cubie. The one flipped edge was therefore moved an odd number of times by the algorithm, while each of the other 11 edges was moved an even number of times. The sum of 11 even numbers and one odd number is always odd, but we established earlier that this sum must be a multiple of 4. An odd number is

a multiple of 4? That is impossible. Therefore, no such algorithm exists.

You have now explored $(3^8 8!)(2^{12} 12!)/12$, the number of cube configurations, which, to a mathematician studying the cube, is just preliminary. To go deeper into the math, you might wonder a common meta-question: "Are there any unanswered math questions in this subject?"

God's Number and Beyond

The original challenge of the cube, of course, was solving it. Ernő Rubik made his first prototype in 1974, and early in the six years it took him to see it mass-produced, he was naturally the first person to ever solve the cube.

When the cube hit toy stores in 1980, some mathematicians had already been experimenting with early versions for a few years. One of them was Dr. David Singmaster, who wrote the famous guide "Notes on Rubik's 'Magic Cube'" and developed a writing method for describing turns of the cube's faces. That notation has become the standard, and it is now known as Singmaster notation.

If this were an article in the 1980s, the labor of explaining Singmaster notation, and using it to guide you through the algorithms of solving the cube, might be worth it. Plenty of articles did just that. But now Youtube tutorials exist,[1] so that practice is obsolete.

The fastest solution times for Rubik's Cube have steadily crept down over the decades. The world record by a human is currently 3.47 seconds.[2] Instrumental to this era of speed cubing was Dr. Jessica Fridrich, who in 1997 developed a method for solving the cube faster than ever. Most of the fastest cube solvers nowadays use some version of the Fridrich method.

As some people sharpened their dexterity, others honed in on the ultimate math questions of Rubik's Cube. No matter how scrambled a cube becomes, how few moves can be applied to solve it? If someone scrambled your cube using 500 moves, it is certainly possible to unscramble it in fewer than 500 moves. But how many fewer?

Thus, the pinnacle of the math in this subject was identified: Is there a magic number that allows us to say, "every scrambled cube can be solved in this many moves [or fewer]"? Thanks to early quips about divine intervention being necessary to glean it with confidence, that number became known as "God's number."

The first major insight on God's number was by Dr. Morwen Thistlethwaite in 1981, who proved that it was at most 52. That means he proved that every scrambled cube can be solved in 52 moves or fewer.

Progress continued through the 1990s and 2000s. Finally, in June 2010, a team of four scientists proved that God's number is 20.[3] That website, which the scientists have maintained ever since, contains the most advanced knowledge about Rubik's Cube to date.

So no matter how scrambled a Rubik's Cube looks, it is always 20 moves away from being solved.

Only small tidbits of math remain unresolved for Rubik's Cube. While God's number is 20, it is unknown exactly how many of the 43,252,003,274,489,856,000 combinations require a whole 20 moves to be solved.

The number of positions that require exactly one move to be solved is 18. That is easy to calculate: There are six faces and three ways to twist each one. How many cubes are exactly two or three moves from being solved is not tough for mathematicians to calculate, but you can imagine that the higher numbers become tricky. The current knowledge goes up to 15; we know exactly how many positions are 15 moves from being solved, but not precisely how many for 16 through 20 moves.

And that is the last math question for the Rubik's Cube. Now you are all caught up until someone answers it. We will let you know when we do.

Notes

1. https://www.youtube.com/results?search_query=how+to+solve+a+rubik%27s+cube.

2. https://www.youtube.com/watch?v=cm5vp4z5l5Y.

3. http://www.cube20.org/.

What Is a Hyperbolic 3-Manifold?

COLIN ADAMS

The simplest example of a hyperbolic manifold is hyperbolic geometry itself, which we describe using the Poincaré disk model. In Figure 1, we see the 2-dimensional version $\mathbb{H}^2$ and the 3-dimensional version $\mathbb{H}^3$, each the interior of a unit 2-disk or 3-disk. In both cases, geodesics are diameters or segments of circles perpendicular to the missing boundary. Notice that for any triangle with geodesic edges, the sum of the angles adds up to less than 180 degrees. This choice of geodesics can be used to determine a corresponding metric, which turns out to have constant sectional curvature -1, justifying the statement that hyperbolic space is negatively curved.

We say that a surface (a 2-manifold) is hyperbolic if it also has a metric of constant sectional curvature -1. We can use this as a definition of a hyperbolic surface, but there are two other helpful ways to think about a hyperbolic surface.

When a surface S has such a metric, we can show that the universal cover of the surface is $\mathbb{H}^2$ and there is a discrete group of fixed-point free isometries Γ of $\mathbb{H}^2$ that act as the covering transformations such that the quotient of $\mathbb{H}^2$ by the action of Γ is the surface.

By choosing a fundamental domain for the group of isometries Γ, we can also think of S as being obtained from a polygon in $\mathbb{H}^2$ with its edges appropriately glued together in pairs by isometries, as in Figure 2. In particular, at each point in S, there is a neighborhood isometric to a neighborhood in $\mathbb{H}^2$. So locally, our surface appears the same as $\mathbb{H}^2$.

Among all topological surfaces, how prevalent are the hyperbolic surfaces? Considering compact orientable surfaces without boundary, only the sphere and the torus are not hyperbolic. All other orientable surfaces are hyperbolic, as in Figure 3. If we throw in nonorientable surfaces, only the projective plane and the Klein bottle are not

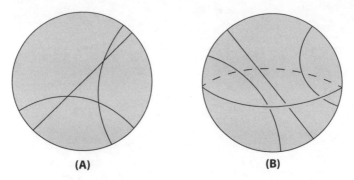

FIGURE 1. The Poincaré disk models of hyperbolic 2-space $\mathbb{H}^2$ and hyperbolic 3-space $\mathbb{H}^3$ with geodesics that are diameters and segments of circles perpendicular to the missing boundary. Image courtesy of Colin Adams.

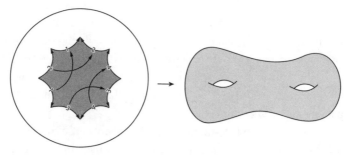

FIGURE 2. Gluing together pairs of edges of a hyperbolic fundamental domain yields the genus two surface. Image courtesy of Colin Adams.

hyperbolic. And if we allow punctures, the only additional surfaces that are not hyperbolic are the once- and twice-punctured sphere and the once-punctured projective plane. So among the infinitude of closed surfaces and closed surfaces with arbitrarily many punctures, all but seven are hyperbolic. So if we want to understand the geometries of surfaces, it's all about the hyperbolic case.

A 3-manifold is a topological space M that is locally 3-dimensional. That is to say, every point has a neighborhood in the space that is homeomorphic to a 3-dimensional ball. For instance, the 3-dimensional spatial universe in which we all live is such a 3-manifold. Another example would be to take 3-space (or the 3-dimensional sphere if we want to begin with a compact space) and remove a knot. Then it is still true that this is a 3-manifold, as every point still has a ball about it that

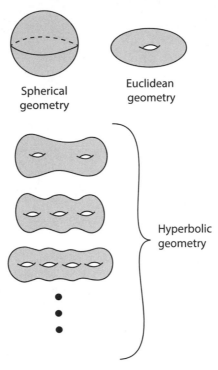

FIGURE 3. A list of closed orientable surfaces and their respective geometries. Image courtesy of Colin Adams.

is 3-dimensional. We just have to pick the ball small enough to avoid the missing knot.

In the 1970s and 1980s, work of William Thurston (1946–2012) and others led to the realization that many 3-manifolds are hyperbolic. Here again, to be hyperbolic just means that there is a metric of constant sectional curvature − 1 or, equivalently, that there is a discrete group of fixed-point free isometries Γ acting on H^3 such that the quotient of the action is M.

A famous example is the figure-eight knot complement. Here, the fundamental domain for the action of the discrete group of isometries is a pair of ideal regular hyperbolic tetrahedra (all angles between faces are $\pi/3$), as in Figure 4.

An ideal hyperbolic tetrahedron is one with geodesic edges and faces such that it is missing its vertices as they sit on the missing boundary of

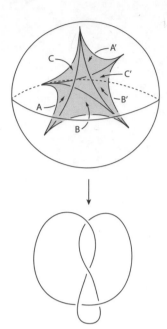

FIGURE 4. A fundamental domain for the figure-eight knot complement constructed from two ideal regular tetrahedra in $\mathbb{H}^3$. Image courtesy of Colin Adams.

$\mathbb{H}^3$. The sum of the volumes of this pair of ideal regular tetrahedra is 2.0298..., a number of interest to number theorists as well as topologists, since it is also related to the value of the Dedekind zeta function at 2. (*See*, for instance, Zagier 1986.) This volume was proved to be the smallest hyperbolic volume of any knot by Cao and Meyerhoff (Cao and Meyerhoff 2001).

Why is it useful for a 3-manifold to be hyperbolic? One extraordinary advantage is the Mostow–Prasad Rigidity Theorem, which says that if you have a finite volume hyperbolic 3-manifold, its hyperbolic structure is completely rigid. All such structures on a given 3-manifold are isometric. In particular, every such 3-manifold has a unique volume associated with it. We have turned floppy topology into rigid geometry.

Compare that to the Euclidean case. We could take a cube and glue opposite faces straight across. This yields the 3-dimensional torus. But we can make it out of a small cube or a big cube, so there is no unique volume associated with it. We could even trade in the cube for a parallelepiped, and we would still have a valid Euclidean structure on the 3-torus.

On the other hand, the figure-eight knot has a hyperbolic complement with volume 2.0298.... So we now have an incredibly effective

FIGURE 5. A torus knot at top and a satellite knot, bottom right. Every other knot must be hyperbolic. Source: Francis 2007. Image courtesy of George Francis.

invariant for distinguishing between 3-manifolds. This was an essential tool used in the classification of the 1,701,936 prime knots through 16 crossings by Hoste et al. in 1998.

In the case of knots, volume is not enough to completely distinguish them for two reasons. First, there are nonhyperbolic knots. Thurston showed that knots fall into three categories: they can be torus knots, satellite knots, or hyperbolic knots. A torus knot is a knot that lives on the surface of an unknotted torus, as in Figure 5, and is determined by how many times it wraps the long and short way around the torus.

A satellite knot is what you might guess, a knot K that orbits another knot K' in the sense that it exists in a neighborhood of K', which is to say a solid torus with K' as core curve. It is a truly marvelous fact that after excluding just these two categories of knots, all other knots are hyperbolic.

Second, although rare for low crossing number, there can be two different knots with the same volume. For instance, the second hyperbolic

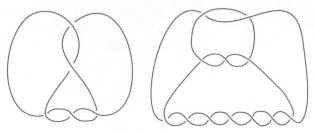

FIGURE 6. There do exist examples of hyperbolic knots with the same volume, such as the pair pictured here. Image courtesy of Colin Adams.

knot 5_2 has the same volume as the 12-crossing $(-2, 3, 7)$-pretzel knot, as in Figure 6.

What we would like is a complete classification of all closed 3-manifolds. This means we would like a way to "list" them all and to decide, given any two, whether or not they are homeomorphic.

In 1982, William Thurston proposed the Geometrization Conjecture. It says that every closed 3-manifold can be cut open along an essential set of tori and spheres into pieces, and after capping off the spheres with balls, each of the components would be 3-manifolds with one of eight specified geometries, one of which is $\mathbb{H}^3$. In 2002/2003, in three preprints posted on the article depository the ArXiv (2002, 2003a,b), Grigori Perelman revolutionized low-dimensional topology by proving the Geometrization Conjecture. (He also proved the Poincaré Conjecture in the process, which was a necessary piece in the proof of the larger Geometrization Conjecture.)

So we would like to determine the manifolds with each of the eight geometries. In fact, the manifolds associated to the seven other geometries have been classified and are well understood. There only remain the manifolds that are hyperbolic. Why have we not succeeded in classifying those? The situation is analogous to what happened with surfaces. This is the richest of the geometries, with the preponderance of the manifolds. It is the mother lode.

Thurston also proposed the Virtual Haken Conjecture (1982), implicit in the work of Waldhausen (1968), that every closed 3-manifold satisfying mild conditions (having infinite fundamental group and no essential spheres) either contains an embedded essential surface or possesses a finite cover that does so, thereby allowing the decomposition along the surface into simpler pieces. The proof of the Geometrization Conjecture allowed for a proof of the Virtual Haken Conjecture for all 3-manifolds except hyperbolic 3-manifolds, which is not such a surprise, since again, this is where the action is. It was this case that Ian Agol completed in 2012 (published 2013), thereby settling this fundamental conjecture. Agol received the $3 million Breakout Prize in mathematics for this and related work.

Research continues forward as we attempt to understand hyperbolic 3-manifolds, their volumes, and other related invariants. This geometric approach to low-dimensional topology has become fundamental to our understanding of 3-manifolds and will continue to play a critical role for years to come.

Additional Reading

Volumes of Hyperbolic Link Complements, Ian Agol, https://www.ias.edu/ideas/2016/agol -hyperbolic-link-complements.

References

Agol, I. (2013). "The Virtual Haken Conjecture." ArXiv 1204.2810 (2012), *Doc. Math.*, 18, 1045–1087.

Cao, C., and Meyerhoff, G. R. (2001). "The Orientable Cusped Hyperbolic 3-Manifold of Minimum Volume." *Inventiones Math.*, 146, 451–478.

Francis, G. K. (2007). *A Topological Picturebook*. Springer Science, New York.

Hoste, J., Thistlethwaite, M., and Weeks, J. (1998). "The First 1,701,936 Knots," *The Mathematical Intelligencer*, 20, 33–48.

Perelman, G. (2002). "The Entropy Formula for the Ricci Flow and Its Geometric Applications." ArXiv 02311519.

Perelman, G. (2003a). "Finite Extinction Times for the Solutions to the Ricci Flow on Certain 3-Manifolds." ArXiv 0307245.

Perelman, G. (2003b). "Ricci Flow with Surgery on 3-Manifolds." ArXiv 0303109.

Thurston, W. (1982). "Three-Dimensional Manifolds, Kleinian Groups and Hyperbolic Geometry." *Bull. Am. Math. Soc.*, 6(3), 357–381.

Waldhausen, F. (1968). "On Irreducible 3-Manifolds Which Are Sufficiently Large." *Ann. Math.*, 87(1), 56–88.

Zagier, D. (1986). "Hyperbolic 3-Manifolds and Special Values of the Dedekind Zeta Function." *Inventiones Math.*, 83, 285–301.

Higher Dimensional Geometries: What Are They Good For?

Boris Odehnal

1. Introduction

Originally, geometry was the science of measuring the land in order to calculate taxes and divide the fertile land. Later on, early cultures, e.g., the Egyptians, Babylonians, and Greeks, began to detach this science from real-world problems and entered the world of abstract two- and three-dimensional phenomena: things that happened in a plane or in the space of our perception, described with a new vocabulary: points, lines, angles, distances, triangles, and many more. Within this period, many well-known elementary geometric results were discovered, and the techniques of proofs were developed. The old Greeks saw geometry rather as a philosophical discipline than as a part of mathematics. It took more than 2,000 years until mathematicians and especially geometers became aware of geometries that do not fit into two or three dimensions. A major breakthrough was H. Grassmann's *Theory of Linear Extensions* [9] in the middle of the nineteenth century. Grassmann's work was maybe not the first attempt, but it successfully provided mathematicians and geometers with techniques that made it possible to describe higher dimensional geometries. During the end of the nineteenth century, a lot of work on higher dimensional geometries was done. In particular, the Italian school, mainly represented by Cremona [6], Veronese [26], Berzolari [1], and Segre (who also worked out important parts of F. Klein's mathematical encyclopedia [24]), began to study algebraic geometries in spaces of dimensions greater than three. At that time, higher dimensional geometries were not common sense to all mathematicians and geometers. Since history repeats itself, a few of them even doubted the existence of such objects. It was comparable to the physicists' concept of the atom, which entered the scientific stage approximately at that time: Notable scientists denied the existence of atoms, using the argument that atoms cannot be seen.

2. A Huge Variety of Higher Dimensional Geometries

2.1. WHAT IS A DIMENSION?

We agree that the *dimension* is a number that counts the number of *degrees of freedom* in a geometrical object. A line is of dimension one. This must not be confused with the number of points on the line. The same is true for planes: They all are of dimension two, no matter whether it is the Euclidean plane, where we can choose Cartesian coordinates (x, y) in order to fix points, or a projective plane, where homogeneous coordinates $x_0 : x_1 : x_2$ are suitable for describing points (still there are only two degrees of freedom, since $x_0 : x_1 : x_2 \sim 1 : x : y$), or a finite (projective) plane, like the ones depicted in Figure 1.

In the Euclidean plane and, more generally speaking, in any Euclidean space, the dimension gives the number of coordinates that are necessary to describe points. It is a useful and powerful result from differential geometry that any differentiable manifold can locally be mapped to a certain real vector space $\mathbb{R}^n$, and thus, a dimension can be assigned to the manifold. Besides the degrees of freedom of a geometrical system and the number of coordinates that are necessary to determine points, there is a more mathematical notion of dimension related to Grassmann's theory of extension. The dimension of a vector space equals the

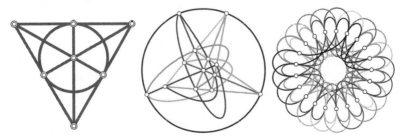

FIGURE 1. Models of projective planes of order two, three, and four. However, these are two-dimensional geometries, even though we can give lists of their points. See also color insert.

number of basis vectors, i.e., a system of linearly independent vectors that allow a unique representation of all elements of a vector space, the vectors. In the case of a vector space, it is assumed that coordinates are real or complex numbers, or taken from an arbitrary field—finite or not. Things become more complicated, and more interesting, once we drop the assumption that coordinates are taken from a field. Geometries over rings are sometimes hard to handle, and models need *more space*, i.e., they are higher dimensional in nature. Moreover, the set of points in geometries over rings can split into different classes, and it is not so easy to compare points [11]. Geometries over finite fields and rings have a lot of applications, especially in physics [10, 13].

However, we shall agree that *higher dimensional geometries and spaces are those with a dimension greater than three.* These are beyond our perception, since we can move forward and backward, to the left and to the right, and up and down in the space where we live. *Time* could be considered a possible fourth dimension, but the perception of time is subject to the individual, and it is not so easy to objectify individual perceptions.

2.2. SOME EXAMPLES OF HIGHER DIMENSIONAL GEOMETRIES

Sometimes the Euclidean unit sphere is called a three-dimensional object. However, this is not true in the strict sense. Two things are mixed up: The sphere itself is two-dimensional, since we need two coordinates to describe points on it: the *latitude* and the *longitude*. However, the sphere is embedded in a three-dimensional space, and it is not possible to embed it into a space of lower dimension without any singularity.

Let us determine dimensions of some known geometries. A three-dimensional (affine) space is usually a space where the manifold of points is three-dimensional, i.e., points are determined by three coordinates (x, y, z). From the equations of planes $ax + by + cz - 1 = 0$ (with a, b, and c taken from some field, not all simultaneously zero), we see that the same space is three-dimensional considered as the space of the planes in it. What about the lines in a 3-space? As indicated in Figure 2, any line l can be uniquely determined by its two intersection points L_1, L_2 with two planes π_1, π_2. Each of these points is determined by two

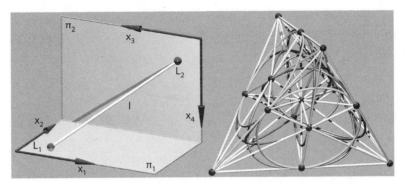

FIGURE 2. Left: Lines can be determined by four coordinates determining the intersection points in two different planes. Right: Though there are only 35 lines in PG(2, 3), the manifold of lines in a 3-space is four-dimensional. See also color insert.

coordinates: $L_1 = (x_1, x_2)$ and $L_2 = (x_3, x_4)$. Since these four numbers $x_1, \ldots, x_4$ can be chosen independently, there are four coordinates that describe a line. Note that the submanifold of lines that meet $\pi_1 \cap \pi_2$ is only of dimension three. Strange to say, but the *space of lines in a three-dimensional space is four-dimensional*.

The geometry of lines plays an essential role in a huge variety of applications (cf. Section 4). Naturally, there is a tremendous amount of literature dealing with line geometry, see [25, 28, 30] and the references therein.

Many higher dimensional geometries are contained within lower dimensional (point) geometries. The geometry of circles in the Euclidean plane is three-dimensional. The center's two coordinates and the radius define a circle.

Conics in a plane can be described by an equation of the form $a_{11}x^2 + 2a_{12}xy + a_{22}y^2 + 2a_{01}x + 2a_{02}y + a_{00} = 0$ with coefficients a_{ij} from some commutative field (cf. [8]). (It is always possible to normalize an equation, i.e., to multiply such that one coefficient becomes unity, without altering the geometric object.) Only the ratio among those six coefficients matters, and so there are five relevant numbers that determine the conic. Thus, the geometry (or manifold) of conics is five-dimensional. A useful tool for the study of conics is due to G. Veronese (see [26]): The six quadratic monomials in the conic's equation can be used as a basis in the projective space of conics. Therefore, each conic can be mapped to a point in a five-dimensional projective space. The manifold of singular conics is called *rank manifold*, whose equation is simply given by $\det(a_{ij}) = 0$.

The reader may convince herself or himself by counting that the space of algebraic curves of degree n in a plane equals $\frac{1}{2}n(n+3)$, including the case of lines and conics. Applying knowledge from basic linear algebra, we find that the manifold of k-dimensional subspaces of a projective space of n dimensions is $(n-k)(k+1)$-dimensional. In any case, one has to think about a proper way of counting. The existence of a vector space model of the geometry in question simplifies the process. However, the dimension of fractals cannot be determined by counting (basis vectors or degrees of freedom). It is to be computed.

3. Model Spaces

We have seen that the geometric objects we are dealing with usually depend on a certain fixed number of constants considered as shape parameters, varying freely within some intervals, or range even in the real or complex number field. It is natural to use these determining constants as coordinates for these objects. The number of these constants equals the dimension of the geometry. In this section, we shall see that model spaces need not be affine, metric, or even projective.

3.1. Various Geometries and Their Models

CIRCLES, SPHERES. Oriented circles, spheres in 3-space, . . . , spheres in an n-dimensional space can be mapped to points in an $n+1$-dimensional affine model space that is usually the Minkowski space $\mathbb{R}^{n,1}$, sometimes referred to as the *cyclographic model*, [4, 7]. The coordinates in the model space are simply the sphere's center plus the radius. Signed radii can be used to express orientations. The pseudo-Euclidean metric in the model space is obtained by transferring the Euclidean *tangential distance* of spheres into the model. The metric allows us to characterize *pairs of spheres* as being in *oriented contact* and can be used to compute oriented intersection angles.

A less natural approach to a point model of the manifold of oriented Euclidean spheres uses (homogeneous) six-tuples $(s_1, \ldots, s_6)$ of real coordinates satisfying the quadratic form $L_4^2 : s_1^2 + s_2^2 + s_3^2 + s_4^2 - s_5^2 - s_6^2 = 0$.

The center and the radius of the sphere can be recovered from this six-tuple as long as it satisfies the quadratic form. The quadric L_2^4 is called *Lie's quadric*. It is contained in a projective 5-space and carries lines as maximal subspaces. This coordinatization of the manifold of spheres is more universal: Points as spheres with radius zero and planes as spheres with infinitely large radius are also described that way, see [2, 4]. Even the polar system of L_2^4 has a geometric meaning: *Conjugacy* with regard to L_2^4 characterizes spheres in *oriented contact*.

The cyclographic model can be linked via a stereographic projection with the *Blaschke model*, i.e., a cylinder model of Euclidean Laguerre geometry (oriented planes, and oriented spheres considered as the envelopes of oriented planes). Blaschke's cylinder is a tangential intersection of Lie's quadric, see [2, 4, 7].

LINES IN 3-SPACE. We have seen that lines in a three-dimensional space can be mapped to points in a four-dimensional model. This naïve approach works well and is even applicable to interpolation problems (cf. [22]), but it is not as universal as the model presented in the following.

It proved useful to describe lines by Plücker coordinates $L = (l_1, l_2, l_3, l_4, l_5, l_6)$ (see [28, 30]). Only those (homogeneous) six-tuples $(l_1, \ldots, l_6)$ that satisfy $M_2^4 : l_1l_4 + l_2l_5 + l_3l_6 = 0$ correspond to lines in 3-space, and, vice versa. This quadratic form is the equation of the four-dimensional model (surface) and describes a quadric M_2^4 in a projective 5-space. It is called *Klein's quadric* or *Plücker's quadric,* and it is also the first non-trivial Grassmannian, and therefore, also denoted by $G_{3,1}$. It is worth mentioning that the maximal subspaces contained in M_2^4 are planes and that there exist two independent three-parameter families of them corresponding to ruled planes and stars of lines in 3-space.

Now, the model space splits into two components: The points on M_2^4 correspond to lines, while the points off M_2^4 correspond to so-called *regular linear line complexes*. The latter are as important in line geometry as the lines, since they are closely related to helical motions. We shall make use of this in Section 4. The polarity with regard to M_2^4 has a geometric meaning: Points *conjugate* with regard to M_2^4 correspond to *intersecting lines* in 3-space. Euclidean specialization of the model can achieve even more, cf. [25, 28, 30].

THE INTERPLAY BETWEEN SPHERES AND LINES. It is obvious that lines and spheres are completely different things. Admittedly, the geometries of both can be modeled within four-dimensional quadrics. However, while M_2^4 carries real planes, L_2^4 carries only real lines. Nonetheless, from the viewpoint of complex projective geometry, the two quadrics M_2^4 and L_2^4 can be transformed into each other by means of a collineation, i.e., a linear transformation in the vector space model. This mapping linking the geometry of lines and the geometry of spheres is called *Lie's line-sphere-mapping* (see [2, 7, 28]). Consequently, there is no difference between lines and spheres, at least in theory.

EUCLIDEAN MOTIONS. Without going too much into detail, we shall recall that Study's quadric S_2^6 serves as a point model for the Euclidean motions in 3-space. This quadric is ruled like L_2^4 and M_2^4 and carries three-dimensional sub-spaces, see [7, 25, 28]. Its equation equals the orthogonality condition of *dual unit quaternions*. Quadrics of the (real) projective type of S_2^6 allow a definition of a so-called *triality* (which generalizes duality), cf. [3, 28].

SUBSPACES OF A PROJECTIVE SPACE. Klein's quadric is a very special version of a Grassmannian. In general, a Grassmannian $G_{n,k}$ is a point model for the set of k-dimensional subspaces in an n-dimensional projective space. The dimensions of the model space are growing rapidly: $G_{n,k}$ spans a projective space of $\binom{n+1}{k+1} - 1$ dimensions and its inner dimension equals $(n - k)(k + 1)$, see [3, 7].

VERONESE VARIETIES AND RATIONAL NORMAL CURVES. As outlined earlier, conics can be studied in the *Veronese model* containing the Veronese surface V_2^2 (all of whose points correspond to conics) and the rank manifold representing the singular conics. The ambient model space is five-dimensional. Clearly, the underlying concept of considering the monomials in the equation of a curve as a basis can be carried over to quadrics, cubics, any algebraic curve, and surface. Here, the *symmetric tensor product* of the underlying vector space builds the algebraic grounding. The study of Veronese manifolds V_1^n, called *rational normal curves*, the n-fold symmetric product of a projective line, is important for the study of rational curves and rational transformations, since any planar rational curve (including Bézier curves) is a projection of a rational normal curve, cf. [3].

SEGRE PRODUCTS, FLAG MANIFOLDS. Geometry models are not restricted to representing only one particular class of object. It is also possible to map combinations of objects to points. It is not at all surprising that the dimension of the model space grows with the complexity

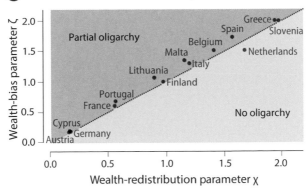

B Phase Transition in Economic Systems

FIGURE 1b from "The Inescapable Casino" (Boghosian)

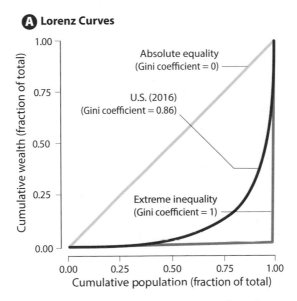

A Lorenz Curves

FIGURE 2a from "The Inescapable Casino" (Boghosian)

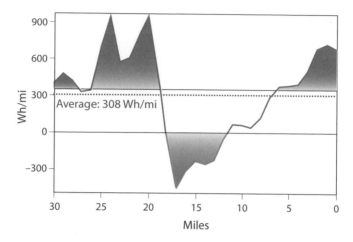

Figure 3 from "Resolving the Fuel Economy Singularity" (Wagon)

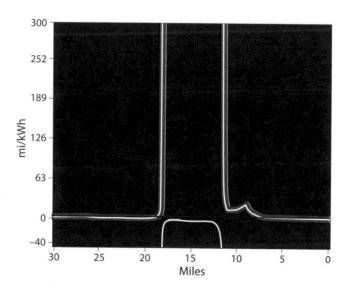

Figure 4 from "Resolving the Fuel Economy Singularity" (Wagon)

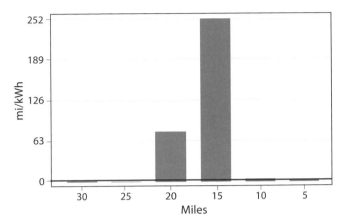

FIGURE 5 from "Resolving the Fuel Economy Singularity" (Wagon)

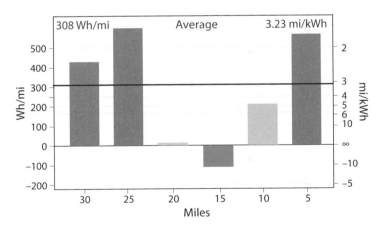

FIGURE 6 from "Resolving the Fuel Economy Singularity" (Wagon)

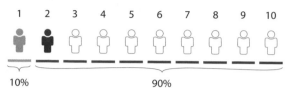

FIGURE 1 from "The Median Voter Theorem" (Veisdal)

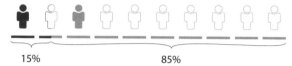

FIGURE 2 from "The Median Voter Theorem" (Veisdal)

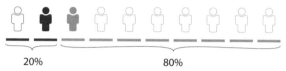

FIGURE 3 from "The Median Voter Theorem" (Veisdal)

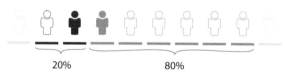

FIGURE 4 from "The Median Voter Theorem" (Veisdal)

FIGURE 5 from "The Median Voter Theorem" (Veisdal)

FIGURE 6 from "The Median Voter Theorem" (Veisdal)

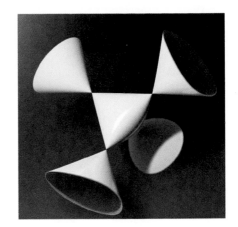

Figure 2 from "The Math That Takes Newton into the Quantum World" (Baez)

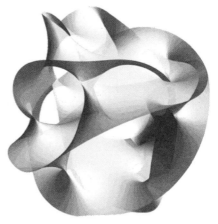

Figure 3 from "The Math That Takes Newton into the Quantum World" (Baez)

Figure 4 from "The Math That Takes Newton into the Quantum World" (Baez)

Boolean Sensitivities

To visualize how sensitive a computer circuit is to bit-flip errors, we can represent its *n* input bits as the coordinates of a corner of an *n*-dimensional cube and color the corner blue if the circuit outputs 1 and red if it outputs 0.

Circuits and Bit-Flips

| OR | Output **1** if <u>any</u> input bit is 1. Output **0** if <u>all</u> input bits are 0. | | AND | Output **1** if <u>all</u> input bits are 1. Output **0** if <u>any</u> input bit is 0. |

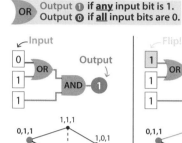

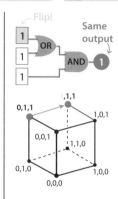

| The output of this simple Boolean function with the input string 011 can be represented as a blue dot at the (0,1,1) corner of this 3D cube. | If you flip the first bit, you move to the blue (1,1,1) corner of the cube. The function is not sensitive to this bit flip. | If instead you flip the third bit, you move to the red (0,1,0) corner of the cube. The function is sensitive to this bit flip. |

Measuring Sensitivity

Once every corner of the cube has been colored according to our Boolean function, the number of sensitive bits for a given input string is captured by the number of connections between its associated corner and corners of the other color. A circuit's overall sensitivity is defined as the largest number of sensitive bits in any input string, so this Boolean function's **sensitivity is 2**.

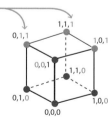

FIGURE 1 from "Decades-Old Computer Science Conjecture Solved in Two Pages" (Klarreich). Infographic created by Lucy Reading-Ikkanda.

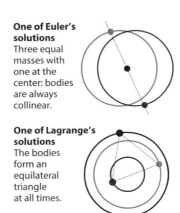

One of Euler's solutions
Three equal masses with one at the center: bodies are always collinear.

One of Lagrange's solutions
The bodies form an equilateral triangle at all times.

FIGURE 2 from "The Three-Body Problem" (Montgomery)

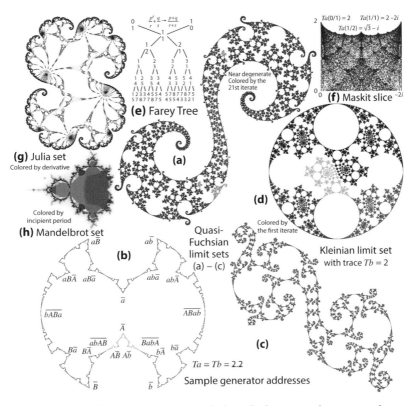

$\frac{p}{r} + \frac{q}{s} \to \frac{p+q}{r+s}$

(e) Farey Tree

Near degenerate
Colored by the 21st iterate

$Ta(0/1) = 2 \quad Ta(1/1) = 2 - 2i$
$Ta(1/2) = \sqrt{3} - i$

(f) Maskit slice $^{-2i}$

(g) Julia set
Colored by derivative

(a)

(d)

Colored by the first iterate

Colored by incipient period
(h) Mandelbrot set

Quasi-Fuchsian limit sets (a) – (c)

(b)
$\overline{aB}$ $\overline{ab}$
$\overline{aB\bar{A}}$ $\overline{aB\bar{a}}$ $\overline{ab\bar{a}}$ $\overline{ab\bar{A}}$
$\bar{a}$
$\overline{bAB a}$ $\overline{ABab}$
$\bar{A}$
$\overline{abAB}$ $\overline{BabA}$
$\overline{Ba}$ $\overline{B\bar{A}}$ $\overline{A\bar{B}}$ $\overline{A\bar{b}}$ $\overline{b\bar{A}}$ $\overline{b\bar{a}}$
$\overline{B}$ $\bar{b}$

Kleinian limit set with trace $Tb = 2$

(c)

$Ta = Tb = 2.2$
Sample generator addresses

FIGURE 1 from "The Intrigues and Delights of Kleinian and Quasi-Fuchsian Limit Sets" (King)

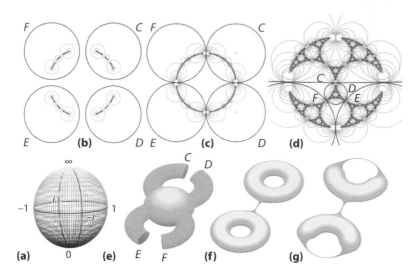

FIGURE 2 from "The Intrigues and Delights of Kleinian and Quasi-Fuchsian Limit Sets" (King)

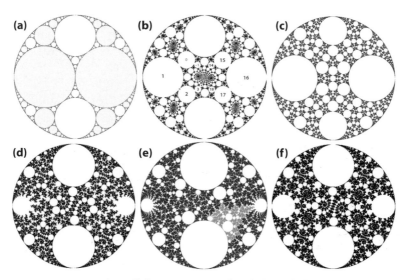

FIGURE 3 from "The Intrigues and Delights of Kleinian and Quasi-Fuchsian Limit Sets" (King)

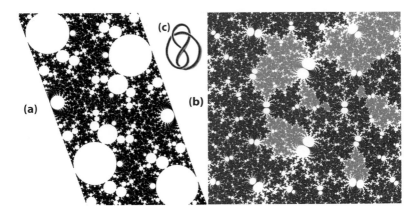

FIGURE 4 from "The Intrigues and Delights of Kleinian and
Quasi-Fuchsian Limit Sets" (King)

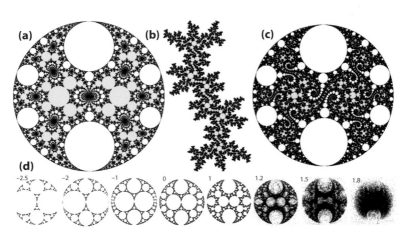

FIGURE 6 from "The Intrigues and Delights of Kleinian and
Quasi-Fuchsian Limit Sets" (King)

FIGURE 1 from "The Amazing Math Inside the Rubik's Cube"
(Linkletter)

FIGURE 2 from "The Amazing Math Inside the Rubik's Cube"
(Linkletter)

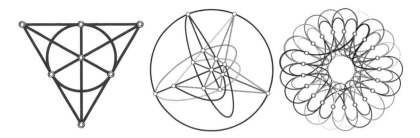

FIGURE 1 from "Higher Dimensional Geometries: What Are They Good For?"
(Odehnal)

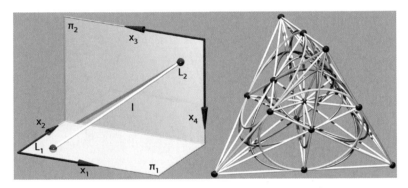

FIGURE 2 from "Higher Dimensional Geometries: What Are They Good For?"
(Odehnal)

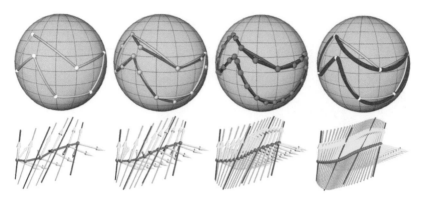

FIGURE 4 from "Higher Dimensional Geometries: What Are They Good For?"
(Odehnal)

FIGURE 5 from "Higher Dimensional Geometries: What Are They Good For?"
(Odehnal)

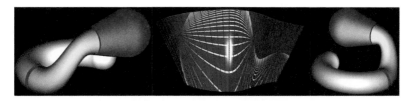

FIGURE 6 from "Higher Dimensional Geometries: What Are They Good For?"
(Odehnal)

FIGURE 7 from "Higher Dimensional Geometries: What Are They Good For?"
(Odehnal)

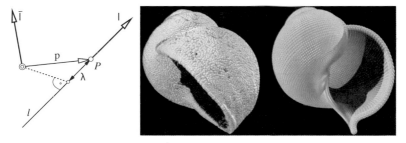

FIGURE 8 from "Higher Dimensional Geometries: What Are They Good For?"
(Odehnal)

FIGURE 9 from "Higher Dimensional Geometries: What Are They Good For?"
(Odehnal)

FIGURE 10 from "Higher Dimensional Geometries: What Are They Good For?"
(Odehnal)

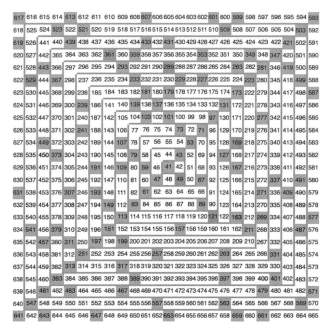

FIGURE 2 from "Who Mourns the Tenth Heegner Number?" (Propp)

How to Multiply Big Numbers Fast

For millennia, it took about n^2 steps of single-digit multiplications to multiply two n-digit numbers. Then in 1960, the Russian mathematician Anatoly Karatsuba proposed a better way.

Traditional Way to Multiply 25 × 63 ————————————————————————

Requires **four** single-digit multiplications and some additions.

STEP (A)	(B)	(C)	(D)	(E)
25	25	25	25	
×63	×63	×63	×63	
1200	15	60	300	1575

1200 **+** 15 **+** 60 **+** 300 **=** 1575

Karatsuba Method for 25 × 63 ————————————————————————

Requires **three** single-digit multiplications **plus some** additions and subtractions.

STEP (A) Break numbers up.	(B) Multiply the tens.	(C) Multiply the ones.	(D) Add the digits.	(E) Multiply the sums.	(F) Subtract B and C from E.	(G) Assemble the numbers.
25 → 2 5	2	5	2 + 5 = 7	7	63	12
63 → 6 3	×6	×3	6 + 3 = 9	×9	− 15	36
	12	15		63	− 12	+ 15
					36	1575

MULTIPLIED SAVINGS: As numbers increase in size, the Karatsuba method can be used repeatedly, breaking large numbers into small pieces to save an increasing number of single-digit multiplications.

Traditional way to multiply 2,531 × 1,467 requires **16 single-digit multiplications.**

2 5 3 1 / 1 4 6 7 **+** 2 5 3 1 / 1 4 6 7 **+** 2 5 3 7 / 1 4 6 7 **+** 2 5 3 1 / 1 4 6 7 **=** 3712977

Karatsuba method to multiply 2,531 × 1,467 requires **9 single-digit multiplications.**

STEP (A)	(B)	(C)	(D)	(E)	(F)	(G)
25 31	25	31	25 + 31 = 56	56	4536	350
14 67	×14	×67	14 + 67 = 81	×81	− 2077	2109
	350	2077		4536	− 350	+ 2077
					2109	3712977

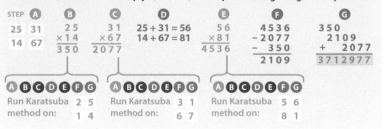

(A)(B)(C)(D)(E)(F)(G)
Run Karatsuba method on: 2 5 / 1 4

(A)(B)(C)(D)(E)(F)(G)
Run Karatsuba method on: 3 1 / 6 7

(A)(B)(C)(D)(E)(F)(G)
Run Karatsuba method on: 5 6 / 8 1

FIGURE 1 from "On Your Mark, Get Set, Multiply" (Honner)

of the underlying objects. Models with the ability to simplify objects consisting of components of various classes of geometric objects can be built using Segre varieties, see [3]. This notion allows us to create models, for example, for the manifold of flags, i.e., a sequence of nested subspaces in some projective space, see [12, 17–19]. The flags need not be complete; some components may be missing: For example, a line element is a *partial flag*. The incidence conditions between the components give rise to equations of flag manifolds.

EXTERIOR ALGEBRAS, CLIFFORD ALGEBRAS. The direct sum over all model spaces that contain the Grassmannians $G_{n,k}$ with $k = 0, \ldots, n$ (with fixed n) is called *exterior algebra* if every summand is considered as a vector space. This 2^n-dimensional vector space can be the algebraic model of a projective space (of dimension $2^n - 1$); it is a model for the set of all subspaces of a projective space of dimension n, cf. [3]. Sometimes, exterior algebras earn the structure of the underlying vector space. This turned out to be useful in kinematics, even in non-Euclidean geometries, see [16].

3.2. WHAT MAKES A MODEL SPACE?

The mere fact that a geometric object can be mapped to a point in some strange high-dimensional space is not enough. The model space itself would be nothing if there is no structure in it. Sometimes, the structuring features come along with the geometries in a natural way; sometimes one has to be creative. As we have seen with lines and spheres, there is a quadratic form in the model space with a geometric meaning. At this point, we note that many of the presented model spaces can be created in a purely synthetic way. In all the aforementioned cases, we always assumed the existence of an algebraic model space, since applications need computations in almost any case. A good model space is easy to handle: It should be affine or projective, of lowest possible dimension, the coordinates should have a geometric meaning, and a metric (a quadratic form) should relate the points in the model. Transformations that act on the manifold of certain geometric objects should be easily transformed to the model space. Preferably, the induced transformations are linear in terms of the coordinates in the model space. As many properties of the underlying geometry as possible should be displayed in a very simple way in the model space.

4. Applications—Benefiting from Model Spaces

4.1. Interpolation with Ruled and Channel Surfaces

In the point models for the set of (oriented) lines/spheres in Euclidean 3-space, we recognize one-parameter families of (oriented) lines/spheres, i.e., *ruled surfaces* or *channel surfaces* as curves on M_2^4 and L_2^4, respectively. So, the geometry of ruled or channel surfaces in a three-dimensional space (whether Euclidean or not) *somehow simplifies* to the geometry of curves on quadrics. The simplification is bought at the costs of more coordinates.

Interpolation techniques and approximation techniques (as shown in Figure 3) that were originally developed for affine planes and spaces (see [15]) can be adapted to arbitrary manifolds. In many cases, the adapted subdivision schemes are combinations of two operations: First, a subdivision scheme in the ambient space of the target manifold, and second, a projection onto the target manifold. In any case, one has to make sure that the projection does not fail and destroy the result. Checking the convergence of a subdivision scheme is not the problem. In the case of ruled surfaces, one can also use ordinary subdivision schemes in order to first refine the striction curve and then refine the spherical image of the rulings. However, this is only one pssibility, cf. [20]. Figure 4 shows the action of an interpolatory scheme on the sphere and applied to a finite sequence of lines (discrete ruled surfaces). Subdivision of the motion of the Sannia frame is obtained as a byproduct, see [20]. Since there is only a small difference between the geometry of oriented

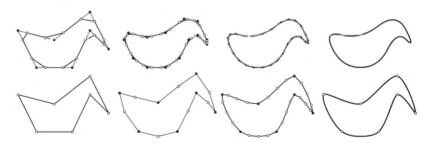

FIGURE 3. Approximating (above) and interpolatory subdivision scheme (below).

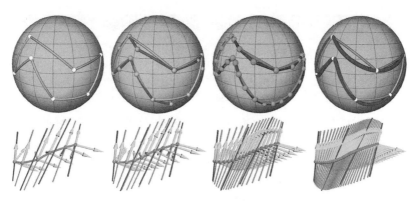

FIGURE 4. Above: Spherical linear interpolation (or slerp) takes place on the sphere. Below: Subdivision of ruled surface data is also suitable for discrete motions (here the Sannia motion), and it uses slerp for the direction of the rulings. See also color insert.

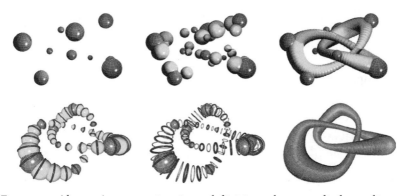

FIGURE 5. Above: An approximating subdivision scheme applied to a discrete channel surface. Below: The set of characteristic circles of a channel surface can also be refined. See also color insert.

lines and oriented spheres in Euclidean 3-space, the algorithms developed for ruled surfaces apply in almost the same way to channel surfaces (Figure 5). Even the characteristic circles on a channel surface are accessible to modified subdivision schemes that take place on a six-dimensional cone-shaped variety, which can be obtained as a projection of the Grassmannian of two Lie quadrics, see [5]. Algebraic techniques like the ones used for G^r or even C^r interpolation of data from curves (as described in [15]) need only some minor modifications in order to

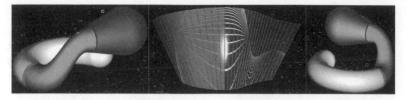

FIGURE 6. Hermite interpolation of ruled and channel surfaces uses a projective 5-space. See also color insert.

apply to interpolation problems with ruled/channel surfaces (cf. [21]). This situation allows us to perform G^r interpolation of data that stems from ruled/channel surface data (Figure 6).

4.2. RECOGNITION AND RECONSTRUCTION OF SURFACES

Surface recognition benefits from line geometry as well as from line element geometry (cf. [14, 18, 23]). It is well known that the normals of helical surfaces (including surfaces of revolution and cylinders) are contained in a *linear line complex*. The Plücker coordinates of the lines of such a three-dimensional submanifold of M_2^4 fulfill a *linear* homogeneous equation. Once a surface is captured by a laser scanner, the point cloud allows an estimation of the surface normals. Fitting linear subspaces to point data in the model space is a simple task, and the computation of the axis and the pitch of the helical motion generating the scanned surface part is straightforward. Figure 7 shows a comparison of the two

FIGURE 7. Left and middle: Scans of the articulate surfaces of the ankle joint. Right: The gliding of the contact surfaces generates a helical motion. See also color insert.

flanks of the human ankle joint. This helped to clarify whether the motion of the human ankle joint is a pure rotation or a helical motion [29]. The comparison of the surfaces was only possible once the motion defined by the flanks was known. For objects composed of many different surfaces, a segmentation is necessary.

An obvious extension of line geometry is called *line element geometry.* It is the geometry of pairs (l, P) where l is a straight line in Euclidean 3-space with a point P on l. Since the lines in Euclidean 3-space can be identified with a subset of M_2^4 and one further parameter is needed in order to fix P on l, we end up with a quadratic cone L_2^5 erected over M_2^4 serving as a point model for the set of (oriented) line elements in Euclidean 3-space (cf. [18]). The coordinates $(\mathbf{l}, \bar{\mathbf{l}}, \kappa) \in \mathbb{R}^7$ of a line element consist of the *direction vector* $\mathbf{l} \in \mathbb{R}^3$ and the *momentum vector* $\bar{\mathbf{l}}, \in \mathbb{R}^3$ of the line $\mathbf{l}$ (cf. [15, 22, 28, 30]) and satisfy $\langle \mathbf{l}, \bar{\mathbf{l}} \rangle = 0$, where $\langle \cdot, \cdot \rangle$ denotes the canonical scalar product of two vectors. The seventh coordinate $\lambda \in \mathbb{R}$ measures the oriented distance from the line's pedal point to the point $P \in l$ (Figure 8, left). A projective version can be found in [19]. Fitting linear subspaces in the geometry of line elements works well. The reconstruction uses the determined generating Euclidean or equiform motion (i.e., a combination of a Euclidean motion and a uniform scaling) to find a profile curve (Figure 8, middle and left). In contrast to line geometry, a wider class of surfaces (including spiral surfaces, i.e., shells of mollusks) can be detected (Figure 9), since the group of Euclidean motions is a subgroup of the group of equiform motions.

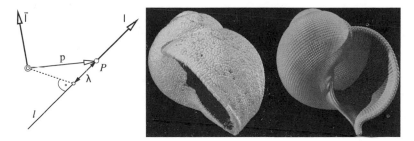

FIGURE 8. Left: The geometric meaning of the coordinates of a line element. Middle: Data from a snail shell can be recognized as a part of a spiral surface. Right: reconstruction. See also color insert.

FIGURE 9. In line element geometry, 11 classes of surfaces can be detected (from top left to bottom right): planes, spheres, spiral cones, cylinders of revolution, spiral cylinders, cones of revolution, spiral surfaces, helical surfaces, surfaces of revolution, generic cylinders, and cones. See also color insert.

FIGURE 10. Left: coordinatization of flags. Right: a Euclidean motion interpolating given poses. The interpolation uses the rationally parametrized manifold of flags. See also color insert.

4.3. INTERPOLATION OF POSES BY SMOOTH MOTIONS

A complete flag (i.e., a plane π containing a line l with an incident point $P \in l$) in Euclidean 3-space determines a Euclidean motion (not necessarily unique). Thus, any model of the geometry of flags in Euclidean 3-space can be used for the design of one-parameter families of Euclidean motions. Either by means of adapted subdivision schemes (as in [20]) or by algebraic techniques (as in [21]). Figure 10 (left) shows how flags can be coordinatized using a vector $(\mathbf{l}, \bar{\mathbf{l}}, \hat{\mathbf{l}}, \lambda) \in \mathbb{R}^{10}$. The equation of the flag manifold in the model space $\mathbb{R}^{10}$ is obtained by the natural constraints to which the flag coordinates are subject: $\langle \mathbf{l}, \bar{\mathbf{l}} \rangle = \langle \mathbf{l}, \hat{\mathbf{l}} \rangle = 0$

(and two vectors are of Euclidean length one, i.e., $\|\mathbf{l}\| = \|\hat{\mathbf{l}}\| = 1$). Combined subdivision techniques like those from [20] apply here as well (Figure 10). A different approach is presented in [27].

5. Conclusion

We have seen that higher dimensional geometries occur frequently and enable us to do interpolation with ruled/channel surfaces, surface reconstruction, motion planning and interpolation, and subdivision in spaces of geometric objects. These are only a few geometries, and we have not treated *shape spaces* that are models for *moving and deformable objects*. Kinematics uses the various model spaces for motion design and the analysis of mechanisms.

References

1. Berzolari, L., Rohn, K. Algebraische Raumkurven und abwickelbare Flächen. Enzykl. Math. Wiss. Bd. 3-2-2a, B.G. Teubner, Leipzig, Germany (1926).

2. Blaschke, W. Vorlesungen über Differentialgeometrie III. Springer, Berlin (1929).

3. Burau, W. Mehrdimensionale und höhere projektive Geometrie. VEB Deutscher Verlag der Wissenschaften, Berlin (1961).

4. Cecil, T.E. Lie Sphere Geometry, 2nd ed. Springer, New York (2008).

5. Coolidge, J.L. A Treatise on the Circle and the Sphere. Clarendon, Oxford, U.K. (1916).

6. Cremona, L. Elemente der Projektiven Geometrie. Verlag Cotta, Stuttgart, Germany (1882).

7. Giering, O. Vorlesungen über Höhere Geometrie. Vieweg, Braunschweig, Germany (1982).

8. Glaeser, G., Stachel, H., Odehnal, B. The Universe of Conics. From the ancient Greeks to 21st century developments. Springer-Verlag, Heidelberg, Germany (2016).

9. Graßmann, H. Die Ausdehnungslehre. Verlag Th. Enslin, Berlin (1862).

10. Havlicek, H., Odehnal, B., Saniga, M. Factor-group-generated polar spaces and (multi-) Qudits. SIGMA Symm. Integrab. Geom. Meth. Appl. 5/098 (2009).

11. Havlicek, H., Kosiorek, J., Odehnal, B. A point model for the free cyclic submodules over ternions. Results Math. **63**, 1071–1078 (2013).

12. Havlicek, H., List, K., Zanella, C. On automorphisms of flag spaces. Linear Multilinear Algebra **50**, 241–251 (2002).

13. Hirschfeld, J.W.P. Projective Geometries over Finite Fields, 2nd ed. Clarendon Press, Oxford, U.K. (1998).

14. Hofer, M., Odehnal, B., Pottmann, H., Steiner, T., Wallner, J. 3D shape recognition and reconstruction based on line element geometry. In: 10th IEEE International Conference Computer Vision, vol. 2, pp. 1532–1538. IEEE Computer Society, 2005.

15. Hoschek, J., Lasser, D. Fundamentals of Computer Aided Geometric Design. A.K. Peters Ltd., Natick, MA (1993).

16. Klawitter, D. Clifford Algebras. Geometric Modelling and Chain Geometries with Application in Kinematics. Ph.D. thesis, TU Dresden, Germany (2015).

17. Odehnal, B. Flags in Euclidean three-space. Math. Pannon. **17**(1), 29–48 (2006).

18. Odehnal, B., Pottmann, H., Wallner, J. Equiform kinematics and the geometry of line elements. Beitr. Algebra Geom. **47**(2), 567–582 (2006).

19. Odehnal, B. Die Linienelemente des P^3. Österreich. Akad. Wiss. math.-naturw. Kl. S.-B. **II**(215), 155–171 (2006).

20. Odehnal, B. Subdivision algorithms for ruled surfaces. J. Geom. Graphics **12**(1), 35–52 (2008).

21. Odehnal, B. Hermite interpolation with ruled and channel surfaces. G – slovenský Časopis pre Geometriu a Grafiku **14**(28), 35–58 (2017).

22. Pottmann, H., Wallner, J. Computational Line Geometry. Springer, Berlin – Heidelberg – New York (2001).

23. Pottmann, H., Hofer, M., Odehnal, B., Wallner, J. Line geometry for 3D shape understanding and reconstruction. In: Pajdla, T., Matas, J. (eds.) Computer vision – ECCV 2004, Part I, vol. 3021 of Lecture Notes in Computer Science, pp. 297–309. Springer, 2004.

24. Segre, C. Mehrdimensionale Räume. Enzykl. Math. Wiss. Bd. 3-2-2a, B.G. Teubner, Leipzig, Germany (1912).

25. Study E. Geometrie der Dynamen. B.G. Teubner, Leipzig, Germany (1903).

26. Veronese, G. Grundzüge der Geometrie von mehreren Dimensionen und mehreren Arten geradliniger Einheiten, in elementarer Form entwickelt. B.G. Teubner, Leipzig, Germany (1894).

27. Wallner, J., Pottmann, H. Intrinsic Subdivision with Smooth Limits for Graphics and Animation. ACM Trans. Graphics **25**/2 (2006), 356–374.

28. Weiss, E.A. Einführung in die Liniengeometrie und Kinematik. B.G. Teubner, Leipzig, Germany (1935).

29. Windisch, G., Odehnal, B., Reimann, R., Anderhuber, F., Stachel, H. Contact areas of the tibiotalar joint. J. Orthopedic Res. **25**(11), 1481–1487 (2007).

30. Zindler, K. Liniengeometrie mit Anwendungen I. II. G.J. Göschen'sche Verlagshandlung, Leipzig, Germany (1906).

Who Mourns the Tenth Heegner Number?

James Propp

There's an episode of *Star Trek: Deep Space Nine* in which space travelers land on a planet peopled by their own descendants. The descendants explain that the travelers will try to leave the planet and fail, accidentally stranding themselves several centuries in the past. Armed with this knowledge, the travelers can try to thwart their destiny—but are they willing to try, if their successful escape would doom their descendants, leaving the travelers with the memory of descendants who, thanks to their escape, never were?

This is science fiction, but it's also math. More specifically, it's proof by contradiction. As Ben Blum-Smith recently wrote on Twitter, "Sufficiently long contradiction proofs *make me sad*! When you stick with the mathematical characters long enough, you start to get attached, and then they disappear, never to have existed in the first place."

This essay is about things that seem to exist but that, when we study them deeply enough, turn out not to exist after all, and about how that disillusionment feels (Figure 1).

Odd perfect numbers are a good example, unless they aren't. A number is *perfect* if it is the sum of its proper divisors. For instance, 6 is perfect because $6 = 1 + 2 + 3$, and 28 is perfect because $28 = 1 + 2 + 4 + 7 + 14$. Nobody's found any odd perfect numbers yet, but we've learned a lot about them; someday we may know so much about them that we'll be able to conclude that there aren't any. Or maybe the dossier of properties odd perfect numbers must satisfy will guide us to one.

This uncertainty is endemic to mathematical research; often we don't know whether our efforts will result in a proof or a counterexample. Douglas Adams wrote that the way to fly is to throw yourself at the ground and miss. In a similar fashion, the way to find a proof is to throw yourself into constructing a counterexample and fail, and the

FIGURE 1. Cartoon by Ben Orlin, author of *Math with Bad Drawings* and *Change Is the Only Constant*.

way to find a counterexample is to throw yourself into constructing a proof and fail.

There are horror stories in math about people who didn't throw themselves hard enough and got thrown by others. There is an apocryphal story about a hapless Ph.D. candidate who wrote a dissertation about a certain class of functions only to find out during his thesis defense that all such functions are constant. That's not the same as finding out that the mathematical objects you've been studying don't exist, but it's almost as bad.

Blum-Smith's tweet was prompted by his reading of I. Martin Isaacs's five-page proof of Burnside's pq-theorem in his book *Finite Group Theory*. Ben sent me an annotated summary, which included his own emotional arc alongside the logic of the proof. Metaphorically speaking, in the middle of the proof the clouds part, and sunlight illuminates a hitherto hidden landscape with marvelous structure. The light gets brighter and brighter. At the end, the seas boil away, the sun explodes, and the planet is obliterated. The End. And the landscape he saw wasn't even there: it was a smudge on his glasses, a flaw in his vision, a consequence of his human inability to perceive the whole truth at once.

My favorite example of something that seemed like it might exist until someone proved that it didn't is the tenth Heegner (HAY-g'nur) number, named after the person who proved that it didn't exist. The proof is too technical to include here, so we will start with a simpler example; historically, it may be the first number that never was.

The Last Prime

It is often said that Euclid proved that there are infinitely many primes, but that is an anachronistic paraphrase. Because the Greeks did not have the modern notion of infinite collections, it would be more accurate to say that Euclid proved that there is no last prime.

The famous argument goes like this: multiply all the primes from the first to the last, obtaining some big number N. The number $N + 1$ is not divisible by *any* of the primes because N is divisible by *all* of the primes, and two consecutive integers cannot have a prime factor in common. So, either $N + 1$ is a prime that we missed, or it is a product of smaller primes that we missed, but either way, the list of primes is incomplete.

See what happened? We assumed there is a last prime, then showed that this assumption contains the seeds of its own destruction; the assumption devours itself, ouroboros-like, until nothing is left but our memory of its self-destruction.

Even if we gave this last prime a name, we probably wouldn't have developed feelings for it; the proof was too short. But when a proof gets long enough and we spend enough time wading through a logical swamp in search of the contradiction, we can, on an emotional level, buy into the fiction we're trying to subvert. And this can be even more true on longer timescales, where the suspense lasts not hours but decades or centuries.

Leonhard Euler

For a long time, we have known that there is no simple, practical formula that, when we plug in n, gives the nth prime. However, there's a great example, due to the eighteenth century mathematician Leonhard Euler, of a simple function that generates a lot of primes: the polynomial $n^2 - n + 41$, which takes on prime values for all n between 1 and 40.

A nice way to visualize this phenomenon is with an *Ulam spiral* (Figure 2). If we put 41 in a box in an infinite square grid and travel outward in a spiral, labeling the boxes with successive integers, then there's a long diagonal stretch of boxes (shown in blue) containing nothing but

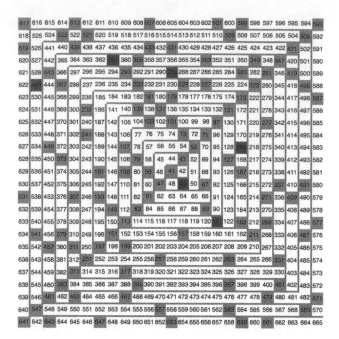

FIGURE 2. An Ulam spiral. See also color insert.

primes. This stretch is so long that the picture is not big enough to contain it! These are the numbers from Euler's function.

Euler may have noticed that there are other numbers k that are "41ish"; that is, for all n between 1 and $k - 1$, $n^2 - n + k$ is prime. Specifically, 1, 2, 3, 5, 11, 17, and 41 have this property, and Euler may have wondered whether there were other numbers with this property.

Carl Friedrich Gauss

Euler's observation took on a new significance in the nineteenth century, when Carl Friedrich Gauss initiated the study of number systems obtained from the field of rational numbers, $\mathbb{Q}$, by throwing in certain irrational numbers. These are called *algebraic number fields* and are often represented by the letter K.

The simplest algebraic number fields are ones we get by inserting $\sqrt{d}$, where d is some rational number that, like -1 or 2, doesn't have a rational square root. This field is written $K = \mathbb{Q}(\sqrt{d})$, and it consists of all numbers of the form $a + b\sqrt{d}$, where a and b are rational numbers. Such a K is called a *real* or *imaginary quadratic number field,* depending on whether $\sqrt{d}$ is real or imaginary. Without loss of generality, we may assume that d is a "square-free" integer; that is, it has no repeated prime factors.

One important quantity associated with a number field K is a positive integer h called the *class number* of K. When $h = 1$, the fundamental theorem of arithmetic holds in K just as it does in $\mathbb{Q}$; that is, every number can be written uniquely as the product of powers of primes. When $h > 1$, the fundamental theorem of arithmetic fails because some of the numbers in K "want" to be factored into primes belonging to a bigger field L, called the *Hilbert class field* of K, and, in a certain sense, L is h times larger than K. (This terminology and description came after Gauss, but the key insight was his.)

In 1801, Gauss observed a curious dichotomy. There are many positive square-free values of d for which the real quadratic field $\mathbb{Q}(\sqrt{d})$ has class number 1 and hence has unique factorization, but he found only a few negative square-free values of d for which the imaginary quadratic field $\mathbb{Q}(\sqrt{d})$ has class number equal to 1, specifically, -1, -2, -3, -7, -11, -19, -43, -67, and -163.

The nine numbers 1, 2, 3, 7, 11, 19, 43, 67, and 163 are now called *Heegner numbers*, although the name is a little misleading; naming this set after Heegner is like naming a species not after the naturalist who discovered them but after the scientist who declared the species extinct!

There is a connection between Heegner numbers and Euler's prime-generating polynomials. To see the connection, discard the first two Heegner numbers and keep the rest: 3, 7, 11, 19, 43, 67, 163. Add 1 to each of them: 4, 8, 12, 20, 44, 68, 164. And now divide each number by 4: 1, 2, 3, 5, 11, 17, 41. These are the k-values for which $n^2 - n + k$ takes on prime values for $n = 1, 2, \ldots, k - 1$.

This is no accident. In 1913, Georg Rabinovitch would show that for $d = 4k - 1$, the field $\mathbb{Q}(\sqrt{-d})$ has class number 1 if and only if $n^2 - n + k$ gives prime values for $n = 1, 2, \ldots, k - 1$.

Gauss conjectured that $\mathbb{Q}(\sqrt{-163})$ was the last of the imaginary quadratic fields with class number 1—that is, as we would put it nowadays, that 163 is the last Heegner number. But he was unable to prove it. This was Gauss's *class number 1 problem,* and it stayed unsolved for more than a century and a half. It is equivalent to the conjecture that Euler's prime-generating polynomial is the last polynomial of its kind.

Charles Hermite

In 1859, the French mathematician Charles Hermite discovered an amazing formula involving Heegner numbers, e, π, and something called the j-function. A consequence of his formula is that when k is 43, 67, or 163, $e^{\pi\sqrt{k}}$ is weirdly close to an integer. Indeed,

$$e^{\pi\sqrt{163}} = 262537412640768743.99999999999925\ldots$$

It is worth stressing how unusual this near miss is: it is rare for irrational numbers to come so close to being whole numbers (aside from numbers like $\sqrt{999999}$, for which the reason for the small difference is obvious).

The math journalist Martin Gardner exploited this near miss in his 1975 April Fools' Day column. He claimed that Srinivasa Ramanujan discovered that $e^{\pi\sqrt{163}} = 262{,}537{,}412{,}640{,}768{,}744$ exactly (Figure 3). This hoax had the unfortunate effect of introducing the memorable caconym "Ramanujan's constant," which has stuck.

If there were a tenth Heegner number, like 163, only bigger, the work of Hermite shows that there would be another numerical coincidence, like $e^{\pi\sqrt{163}}$ but even closer to an integer. And if there were an 11th Heegner number, there would be an even more stupefyingly close coincidence.

Hermite (as far as I'm aware) did not speculate about whether there were more numbers of this kind, but I'm pretty sure that if he gave the matter any thought, he would have agreed with Gauss that there probably aren't any. Yet, for all Hermite and his contemporaries could

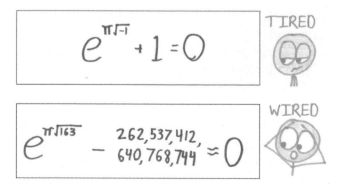

FIGURE 3. The number $e^{\pi\sqrt{163}}$ is almost an integer. Cartoon by Ben Orlin.

prove, there might be infinitely many more such numbers, each giving a more amazing near miss than the one before.

Hans Heilbronn and Edward Linfoot

In 1934, Hans Heilbronn proved a conjecture of Gauss's (called the *class number problem*, not to be confused with the class number 1 problem) that for each positive integer *h*, there are only finitely many imaginary quadratic fields *K* with class number *h*.

This result had implications for Gauss's class number 1 problem; it showed that there are only finitely many imaginary quadratic fields with class number 1. That is, Heilbronn's work showed that even if Gauss was wrong and 163 wasn't the last Heegner number, there *was* a last Heegner number; that is, there were at most finitely many more that Gauss hadn't known about. Maybe none, maybe one, or maybe two or more?

That ambiguity didn't last long. Working with Edward Linfoot in that same year, Heilbronn proved that there is at most one more Heegner number beyond the nine that Gauss and Hermite knew about.

The proviso "at most one" invites us to reify the tenth Heegner number in a way that Heilbronn's earlier result does not. This says more about the human mind than it does about mathematics. We crave concreteness, and "at most one" offers the prospect of specificity, though with no promises.

Kurt Heegner

Although he is largely unknown to nonmathematicians (and to most mathematicians outside of the field of number theory), Kurt Heegner was one of the greatest amateur mathematicians of the twentieth century. He was a radio engineer who pursued number theory as a hobby, and his greatest achievement was his 1952 solution of Gauss's class number 1 problem.

Heegner's proof was regarded as incomplete when he published it, and it was not studied by many number theorists in his lifetime. In the 1960s, Alan Baker and Harold Stark found proofs of their own. When Stark read Heegner's proof, he saw that it was essentially valid and that the gaps in the proof were easily filled; Heegner most likely could have supplied those missing steps had anyone asked him to.

In any case, for hundreds of years, the question of whether there was a tenth Heegner number was open, and for 20 or 30 years, the question was lent extra piquancy by the knowledge that there was no 11th. The theorem of Baker, Heegner, and Stark settled the matter. The ninth Heegner number is the last of its kind; there is no tenth.

The phrase "the tenth Heegner number" cropped up only as a result of historical happenstance. As Stark pointed out in his 1969 paper, if Heinrich Martin Weber had taken a closer look at results in his own 1895 treatise *Lehrbuch der Algebra*, he might have solved the class number 1 problem. And if Weber had found a complete solution of the class number 1 problem before Heilbronn and Linfoot could find a partial one, then the prospect of a tenth and final Heegner number would never have achieved the level of intrigue that it possessed for three decades.

If other intelligent minds exist in the universe and they do mathematics, then there are likely to be satisfying similarities and striking differences. If aliens care about prime numbers, they are bound to discover that there are infinitely many. If they care about number fields and class numbers, they will care about imaginary quadratic fields with class number 1. They will discover Heegner numbers, and they're bound to conclude that there are only nine. But will they ever be at a stage of knowing that there are at least nine and at most ten? It seems to me unlikely that the twists and turns of other planets' mathematical histories will reproduce this particular circumstance. So, if anyone is going to mourn the tenth Heegner number, it will have to be us.

Acknowledgments

The author thanks John Baez, Peter Blauner, Ben Blum-Smith, Veit Elser, Sandi Gubin, Keith Lynch, Kerry Mitchell, Ken Ono, Ben Orlin, and Evan Romer. See his blog for an extended discussion of this topic (bit.ly/Heegner).

On Your Mark, Get Set, Multiply

PATRICK HONNER

This summer, battle lines[1] were drawn over a simple math problem: $8 \div 2(2 + 2) = ?$ If you divide 8 by 2 first, you get 16, but if you multiply 2 by $(2 + 2)$ first, you get 1. So, which answer is right? The conflict grew so heated that it made the pages of *The New York Times*.[2] And as the comments section shows, even a professional mathematician weighing in on the matter was not enough to bring the two sides together.

The problem here is simply how we interpret the division symbol. Does $\div$ mean divide by the one number right after it or by everything after it? This is not much of a concern for most mathematicians, as they do not use this symbol very often. Ask them to solve this problem and they'll probably just make the whole thing into a multiplication problem: Once you choose to write it as either

$$8 \times \frac{1}{2(2+2)} \quad \text{or} \quad 8 \times 1/2(2+2)$$

the ambiguity is gone and the answer is clear. As a multiplication question, this is not particularly interesting.

But one multiplication question mathematicians do find interesting may surprise you: What is the best way to multiply?

Suppose you were asked to multiply 25 and 63. If you are like most people, you would probably reach for a calculator. But if you could not find one, you would probably use the standard algorithm you learned in elementary school, multiplying each digit from one number with each digit from the other and then adding up the products:

$$
\begin{array}{r}
25 \\
\times 63 \\
\hline
15 \\[6pt]
60 \\
300 \\
+1200 \\
\hline
1575
\end{array}
$$

If you are comfortable doing mental math, you might take another approach using the distributive property to make it easier to calculate the answer in your head:

$$25 \times 63 = 25 \times (60 + 3) = 25 \times 60 + 25 \times 3 = 1500 + 75 = 1575$$

Both give us the correct answer, but is one way better than the other? Perhaps it is just personal choice. But there is at least one way to objectively compare multiplication methods: Efficiency. While efficiency can mean different things to different people in different contexts, let's think about it from a computer's perspective. How long would it take a computer to perform the multiplication?

Evaluating the efficiency of computer algorithms is complex, but we take a simple approach that relies on two assumptions. First, the multiplication of two large numbers can always be broken down into a bunch of small multiplications and additions. And second, the way most computers are designed, small additions can be performed much faster than small multiplications. So when it comes to measuring the efficiency of a multiplication algorithm, it is the small multiplications that concern us most.

Let's revisit our example, 25×63, with efficiency in mind. In order to compute 25×63 using the standard algorithm, we had to perform four small multiplications: 3×5, 3×2, 6×5, and 6×2. The small multiplications gave us our rows of 15, 60, 300, and 1,200, which added up to 1,575. It is important to note that when we apply the standard algorithm, 3×2 is really $3 \times 2 \times 10 = 60$, 6×5 is really $6 \times 10 \times 5 = 300$, and 6×2 is really $6 \times 10 \times 2 \times 10 = 1,200$. However, we do not count multiplying by 10 against the efficiency of our algorithm, as it is just as easy for computers to multiply by 10 as it is for us—we just shift the digits over. The same is true for multiplying by 100, 1,000, and so on.

Thus, our execution of the standard algorithm to compute 25×63 requires four small multiplications and some additions. What about the distributive approach?

$$25 \times 63 = 25 \times (60 + 3) = 25 \times 60 + 25 \times 3 = 1500 + 75 = 1575$$

In computing 25×60, we have to multiply 6 by both 2 and 5, and in computing 25×3, we have to multiply 3 by both 2 and 5. That's still four small multiplications. Since each method requires four small multiplications, by our simple measure the two methods are roughly equivalent in terms of efficiency. So it should not come as a surprise that the standard algorithm is really just an application of the distributive property. Let's see why.

Think about multiplying two arbitrary two-digit numbers 'AB' and 'CD.' Here I use single quotes to denote a number described by its digits: 'AB' is the number whose tens digit is A and whose ones digit is B. In other words, 'AB' is the number $10A + B$. So, if 'AB' is the number 25, $A = 2$ and $B = 5$. And if 'CD' is the number 63, $C = 6$ and $D = 3$.

In order to find the product of 'AB' and 'CD', we need to multiply the two numbers $10A + B$ and $10C + D$. We can do this using the distributive property twice:

$$(10A + B) \times (10C + D) = 100(A \times C) + 10(A \times D) + 10(B \times C) + B \times D.$$

Here's how it looks if we plug in our original numbers:

$$25 \times 63 = (10 \times 2 + 5) \times (10 \times 6 + 3)$$
$$= 100(2 \times 6) + 10(2 \times 3) + 10(5 \times 6) + 5 \times 3$$
$$= 1200 + 60 + 300 + 15$$
$$= 1575$$

Notice that all the components of our application of the standard algorithm are there, just organized differently. And in terms of efficiency, we see exactly the same small multiplications performed in both: 3×5, 3×2, 6×5, and 6×2. These two methods are doing essentially the same thing. Either way, we have to multiply $A \times C$, $A \times D$, $B \times C$, and $B \times D$. These four small multiplications appear to set a limit for how efficient multiplication can be.

But in 1960, the Russian mathematician Anatoly Karatsuba found a new limit, using his own application of the distributive law to find a more efficient way to multiply. Karatsuba noticed that all four small multiplications required to compute the product of '*AB*' and '*CD*' appear when we multiply the sums of their digits, $A + B$ and $C + D$:

$$(A + B) \times (C + D) = A \times C + A \times D + B \times C + B \times D.$$

The sum of those four small multiplications is not exactly what we want, but Karatsuba knew he could work with it. Let's see what he did.

In the Karatsuba method for computing '*AB*' $\times$ '*CD*,' we first perform the two small multiplications $A \times C$ and $B \times D$. These are essentially the hundreds digit ($A \times C$) and the ones digit ($B \times D$) of our final answer. (There may be some carrying, but remember, small additions are much faster than small multiplications.) A negligible multiplication by 100 gets us two of the four terms we are looking for:

$$100(A \times C) + B \times D$$

To complete the entire multiplication problem, we want:

$$100(A \times C) + 10(A \times D) + 10(B \times C) + B \times D.$$

So we are only missing $10(A \times D) + 10(B \times C)$. Here's where Karatsuba's clever trick comes in. We can rearrange the product of the sums of digits

$$(A + B) \times (C + D) = A \times C + A \times D + B \times C + B \times D$$

to get

$$(A + B) \times (C + D) - A \times C - B \times D = A \times D + B \times C.$$

We need $10(A \times D) + 10(B \times C)$, which is the same as $10(A \times D + B \times C)$, so we can just multiply both sides of the above equation by 10 to get:

$$10((A + B) \times (C + D) - A \times C - B \times D) = 10(A \times D + B \times C).$$

At first glance this does not seem like an improvement. We have turned something fairly simple with only two small multiplications, $10(A \times D + B \times C)$, into something that has three multiplications, $10((A + B) \times (C + D) - A \times C - B \times D)$. But Karatsuba's ultimate goal was not to make things look nicer. It was to make multiplication more efficient. The secret to Karatsuba's method is that we do not need to compute $A \times C$ and $B \times D$ again: Those were the first two things we multiplied. We already know them!

We are really replacing the two multiplications $A \times D$ and $B \times C$ with two multiplications we have already performed—$A \times C$ and $B \times D$—and one new small multiplication, $(A + B) \times (C + D)$. (Note: the infographic "How to Multiply Big Numbers Fast" can also be found in the special insert with color figures)

How to Multiply Big Numbers Fast

For millennia, it took about n^2 steps of single-digit multiplications to multiply two n-digit numbers. Then in 1960, the Russian mathematician Anatoly Karatsuba proposed a better way.

Traditional Way to Multiply 25 × 63 ———————————————————

Requires **four** single-digit multiplications and some additions.

STEP Ⓐ	Ⓑ	Ⓒ	Ⓓ	
25	25	25	25	Ⓔ
×63	×63	×63	×63	
1200 ✛	15 ✛	60 ✛	300 ═	1575

Karatsuba Method for 25 × 63 ———————————————————

Requires **three** single-digit multiplications plus some additions and subtractions.

STEP Ⓐ Break numbers up.	Ⓑ Multiply the tens.	Ⓒ Multiply the ones.	Ⓓ Add the digits.	Ⓔ Multiply the sums.	Ⓕ Subtract B and C from E.	Ⓖ Assemble the numbers.
25 → 2 5	2	5	2 + 5 = 7	7	63	12
63 → 6 3	×6	×3	6 + 3 = 9	×9	− 15	36
	12	15		63	− 12	+ 15
					36	1575

MULTIPLIED SAVINGS: As numbers increase in size, the Karatsuba method can be used repeatedly, breaking large numbers into small pieces to save an increasing number of single-digit multiplications.

Traditional way to multiply 2,531 × 1,467 requires 16 single-digit multiplications.

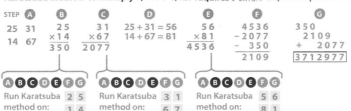

2 5 3 1	2 5 3 1	2 5 3 7	2 5 3 1		
1 4 6 7 ✛	1 4 6 7 ✛	1 4 6 7 ✛	1 4 6 7 ═		3712977

Karatsuba method to multiply 2,531 × 1,467 requires 9 single-digit multiplications.

STEP Ⓐ	Ⓑ	Ⓒ	Ⓓ	Ⓔ	Ⓕ	Ⓖ
25 31	25	31	25 + 31 = 56	56	4536	350
14 67	×14	×67	14 + 67 = 81	×81	− 2077	2109
	350	2077		4536	− 350	+ 2077
					2109	3712977

ⒶⒷⒸⒹⒺⒻⒼ	ⒶⒷⒸⒹⒺⒻⒼ	ⒶⒷⒸⒹⒺⒻⒼ
Run Karatsuba 2 5 method on: 1 4	Run Karatsuba 3 1 method on: 6 7	Run Karatsuba 5 6 method on: 8 1

This means we can compute 'AB' × 'CD' using only three small multiplications: $A \times C$, $B \times D$, and $(A + B) \times (C + D)$.

Karatsuba's method is not easy, and you certainly would not use it to multiply 25 by 63 just to save one small multiplication (*See also* color pages). The method requires more additions, and $(A + B) \times (C + D)$ might not even seem all that small. (Though if you think about it, the largest value of $A + B$ or $C + D$ cannot be much bigger than a single-digit number). The important thing is that in the context of multiplying very large numbers, as when computers use mathematical techniques to encrypt and decrypt secret messages and sensitive data, these small trade-offs add up to big gains in speed.

For example, suppose we wanted to multiply two four-digit numbers like 1,234 and 5,678. Traditional multiplication techniques would require multiplying each digit of one number by each digit of the other, for a total of $4 \times 4 = 16$ small multiplications. But a simple application of Karatsuba's method can reduce that: By thinking of 1,234 as $12 \times 100 + 34$ and 5,678 as $56 \times 100 + 78$ and using the distributive property, we see that:

$$1234 \times 5678 = (12 \times 100 + 34) \times (56 \times 100 + 78)$$
$$= 12 \times 56 \times 10,000 + 12 \times 78 \times 100 + 34 \times 56 \times 100 + 34 \times 78$$

This requires four products of pairs of two-digit numbers, which, at four small multiplications each, give us our 16 multiplications. But by using Karatsuba's method four times, we could reduce those four products to three small multiplications each, for a total of 12. And this 25% improvement is just the beginning. Further clever applications of Karatsuba's method can reduce the number of multiplications even more, and the savings grow as the numbers grow larger.

When multiplying two n-digit numbers together using traditional methods, we would expect to perform $n \times n = n^2$ small multiplications— each digit of the first number times each digit of the second number. But full implementations of Karatsuba's algorithm require only around $n^{1.58}$ small multiplications. This makes a huge difference as the numbers grow larger. Multiplying two 10-digit numbers using traditional methods requires $10 \times 10 = 10^2 = 100$ small multiplications, but only around $10^{1.58} \approx 38$ using Karatsuba's method. That's a 62% decrease. And for two 100-digit numbers, the savings are even greater: $100^2 = 10,000$ versus $100^{1.58} \approx 1,445$, an 85% difference!

You would not use this algorithm when calculating a tip, but when it comes to multiplying large numbers, Karatsuba's method was a big advance. And once Karatsuba opened the door to faster multiplication in 1960, mathematicians have been setting new speed records ever since using advanced techniques like the Fourier transform. This method turns problems about multiplying numbers into problems about multiplying polynomials,[3] where there are surprising shortcuts that create even faster algorithms—ones that computers still use today. These improvements culminated earlier this year, when two researchers verified a nearly 50-year-old conjecture[4] about the maximum efficiency of multiplication methods, finally settling the question about the fastest way to multiply.

But the question of how you should multiply is still open. Know your algorithms, but do what works best for you. And remember, multiplication is not a race—unless you are trying to set a new speed record.

Notes

1. https://twitter.com/kmgelic/status/1155598050959745026?ref_src=twsrc%5Etfw%7Ctwcamp%5Etweetembed%7Ctwterm%5E1155598050959745026&ref_url=https%3A%2F%2Fwww.nytimes.com%2F2019%2F08%2F02%2Fscience%2Fmath-equation-pedmas-bedmas-bedmas.html.

2. https://www.nytimes.com/2019/08/02/science/math-equation-pedmas-bemdas-bedmas.html.

3. https://www.quantamagazine.org/tag/polynomials/.

4. https://www.quantamagazine.org/mathematicians-discover-the-perfect-way-to-multiply-20190411/.

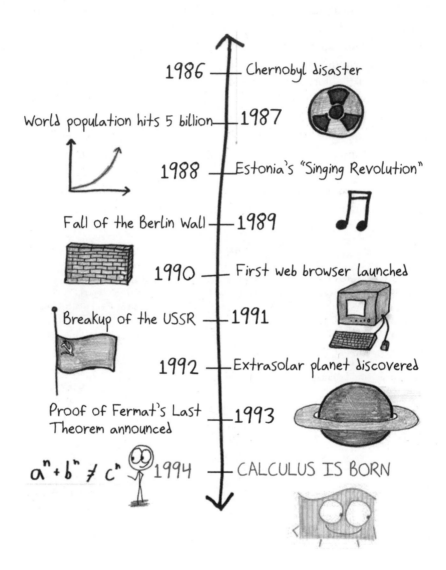

1986 — Chernobyl disaster

World population hits 5 billion — 1987

1988 — Estonia's "Singing Revolution"

Fall of the Berlin Wall — 1989

1990 — First web browser launched

Breakup of the USSR — 1991

1992 — Extrasolar planet discovered

Proof of Fermat's Last Theorem announced — 1993

$a^n + b^n \neq c^n$ 1994 — CALCULUS IS BORN

1994, the Year Calculus Was Born

BEN ORLIN

In February of the 1994th year of the Common Era, the medical journal *Diabetes Care* published an article by researcher Mary Tai. Its title: "A Mathematical Model for the Determination of Total Area under Glucose Tolerance and Other Metabolic Curves."

Sensationalist clickbait, I know, but bear with me.

Whenever you eat food, sugar enters your bloodstream. Your body can make glucose out of anything, even spinach or steak (which is why the "Orlin Diet" skips the middleman and prescribes only cinnamon rolls). Whatever the meal, your blood-sugar level rises and then, over time, returns to normal. Key health questions: How high does it rise? How fast does it fall? And, most of all, what trajectory does it follow?

The "glycemic response" is not just a peak or a duration; it's a whole story, the aggregate of infinite tiny moments. What doctors want to know is the area under the curve.

Glucose Level Over Time

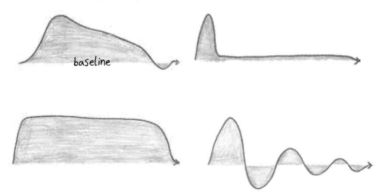

Alas, they can't just invoke the fundamental theorem of calculus. That's for curves defined by tidy formulas, not ones born from a game of connect-the-dots with empirical data. For such messy realities, you need approximate methods.

That's where Tai's paper comes in. "In Tai's Model," it explains, "the total area under a curve is computed by dividing the area . . . into small segments (rectangles and triangles) whose areas can be accurately calculated from their respective geometrical formulas."

TAI'S METHOD

GLUCOSE

Total Area = 240

Tai writes that "other formulas tend to under- or overestimate the total area under a metabolic curve by a large margin." Her method, by contrast, appears accurate to within 0.4%. It's clever geometric stuff, except for one tiny criticism.

This is Calc 101.

Mathematicians have known for centuries that, when it comes to practical approximations, there are better methods than Riemann's skyline of rectangles. In particular, you can identify a series of points along your curve, then connect the dots with straight lines. This creates a procession of long, skinny trapezoids.

TRAPEZOID METHOD

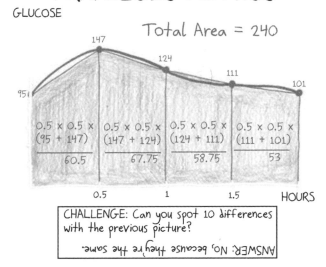

GLUCOSE

Total Area = 240

147

124

111

95

101

0.5 × 0.5 × (95 + 147) / 60.5

0.5 × 0.5 × (147 + 124) / 67.75

0.5 × 0.5 × (124 + 111) / 58.75

0.5 × 0.5 × (111 + 101) / 53

0.5   1   1.5   HOURS

CHALLENGE: Can you spot 10 differences with the previous picture?

ANSWER: No, because they're the same.

Forget 1994. This method wasn't new in 1694, or in 94 BCE. Ancient Babylonians employed it to calculate the distance traveled by the planet Jupiter. Tai had written, and a referee had approved, and *Diabetes Care* had published, a piece of work millennia old, one that a gung-ho undergraduate could do for homework. All as if it were novel.

Mathematicians had a field day.

Event #1: Headshaking. "Tai proposed a simple, well-known formula exaggerately [*sic*] as her own mathematical model," wrote one critic in a letter to *Diabetes Care*, "and presented it in a circumstantial and faulty way."

Event #2: Sneering. "Extreme ignorance of mathematics," one online commenter remarked. "This is hilarious," wrote several others.

Event #3: Reconciliation. "The lesson here is that calculating areas under the curve is deceptively difficult," wrote a diabetes researcher whose earlier paper Tai had criticized (based on an incorrect understanding, it turns out). The letter ended on a conciliatory note: "I fear I may be responsible for contributing to the confusion."

Event #4: Teachable Moments. Two mathematicians pushed back on Tai's insistence that her formula was not about trapezoids, but about triangles and rectangles. They even drew a picture for her: "As is evident in the following figure . . . the small triangle and the contiguous rectangle form a trapezoid."

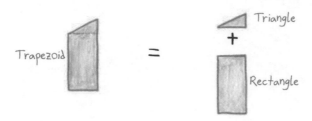

Event #5: Soul-Searching. "As a smug physicist myself," wrote one commenter, in response to a fun-poking blog post, "I did find this funny, but I can't help thinking that this post makes us look worse than them. . . . I'm sure you can find plenty of physicists saying spectacularly naïve things about medicine or economics."

For what it's worth, mathematical researchers have been known to reinvent the wheel, too. During his graduate studies, Alexander Grothendieck rebuilt the Lebesgue integral for himself, not realizing he was replicating old work.

As Tai tells it, she wasn't trying to glorify her method. "I never thought of publishing the model as a great discovery or accomplishment," she wrote. But colleagues "began using it and . . . because the investigators were unable to cite an unpublished work, I submitted it for publication at their requests." She was just trying to share her thinking, to facilitate further inquiry.

Alas: In academia, publishing isn't just about sharing information. It's more than a way to say *Here's a thing I know.* It's also a scoreboard, a declaration of *Hey, here's a thing I know because I discovered it, so please esteem me highly. Thank you and good night.*

This publication system has its flaws. "Our basic and most important unit of discourse is a research article," writes mathematician Izabella

Łaba. "This is a fairly large unit: effectively, we are required to have a new, interesting and significant research result before we are allowed to contribute anything at all."

Łaba likens this to an economy where the smallest denomination is the $20 bill. In such a world, how can anybody run a bake sale? Either you've got to coax folks into buying $20 worth of muffins, or you've got to give them away for free. Tai chose to charge $20, but neither option is much good. "We should have smaller bills in circulation," writes Łaba. "It should be possible to make smaller contributions—on a scale of, say, a substantive blog comment."

What Academia Has

What Academia Needs

Integrals aren't just for mathematicians. Hydrologists use them to estimate the flow of contaminants through groundwater; bioengineers, to test out theories of lung mechanics; Russian novelists, as a metaphor for a democratic vision of history; economists, to analyze a society's income distribution, its deviation from a state of perfect equality. Integrals belong to diabetes researchers, to mechanics, to Leo Tolstoy, to anyone and everyone who seeks the area under a curve, the infinite sum of infinitesimal pieces. The integral is a hammer in a world full of nails, and it doesn't just belong to hammer makers.

But calculus teachers, like your sheepish and regretful author, can slip up. We emphasize the antiderivative approach—which works only in theoretical settings, where we possess an explicit formula for the curve. That privileges philosophy over practice, abstraction over empirics.

It's also outdated. "Numerical analysis has grown into one of the largest branches of mathematics," writes Professor Lloyd N. Trefethen. "Most of the algorithms that make this possible were invented since 1950." The trapezoid rule, alas, is not among them. But calculus, for all its antiquity, remains a growing field, even in the years since 1994.

Gauss's Computation of the Easter Date

Donald Teets

In the article "Berechnung des Osterfestes (Calculation of the Easter Date)" [4] that appeared in August 1800, Carl Friedrich Gauss offers an extraordinarily simple set of arithmetic rules for calculating the Easter date in any given year. He illustrates it with an example, concluding that "so it is, for example, for the year 4763 Easter is . . . the 7th of April. . . ." This article takes a close look at Gauss's algorithm and offers a glimpse at the history surrounding it.

There is a vast literature on the subject of Easter and the Easter date (see the references in [1], for example), so why another paper on this topic? First and foremost, this paper is *pure Gauss*: *his* method and *his* comments, offering readers a glimpse into his original work. Second, Gauss's algorithm, and variations thereof, are widely reproduced in the literature of the subject just as Gauss presented it; that is, with little or no explanation whatsoever. (For example, see the Easter date algorithm in the U.S. Naval Observatory's *Explanatory Supplement to the Astronomical Almanac* [9].) The 10 modular arithmetic formulas appearing in the algorithm are simple to implement, yet so inscrutable that they beg further explanation: Where do they come from, and why does the algorithm work? Third, frequent reference is made to an error in Gauss's algorithm, suggesting that his method should be discounted entirely. We shall see that Gauss is undeserving of much of the criticism surrounding this error.

All in all, despite the immense number of papers on the subject of the Easter date that have appeared through the centuries, the elegance and simplicity of Gauss's algorithm are not widely understood. And surprisingly, this very popular topic seems never to have been addressed in a century of Mathematical Association of America journals.

Gauss was 23 years old and struggling to find an appointment to support himself when "Berechnung des Osterfestes" was published. His biographer G. W. Dunnington [3] writes that "according to his own story his mother could not tell him the exact day on which he was born; she only knew that it was a Wednesday, eight days before Ascension Day. That started him on his search for the formula." (Note: Ascension Day is the Thursday 40 days after Easter, counting the latter as day one. In Gauss's birth year 1777, Easter fell on March 30, Ascension Day was Thursday, May 8, so Gauss calculated his birthday as April 30.)

Gauss achieved fame in the fields of number theory and astronomy, and the Easter date computation is a mix of the two. But each of these appears in a trivial way; the real complexity of the problem stems from the byzantine rules upon which the Easter date is based. In Gauss's own words, "the intention of this article is not to discuss the standard method of computing the Easter date, which one can find in any manual of mathematical chronology, and which is easy enough, once one knows the meaning and practice of the usual vocabulary of the profession, *golden number, epact, Easter limit, solar cycle,* and *Dominical letter,* and has the necessary assistance tables available; rather than this task, to give a means of help, independent and clear, a pure analytical solution based on the simplest calculations and operations."

Definition of Easter

A popular definition of Easter is "the Sunday after the first full moon on or after the vernal equinox." But this definition comes with certain difficulties; for example, a full moon may occur just before midnight in one time zone and just after in another, thus on different days. In the Catholic Encyclopedia [7], we find the following: "Seeing, therefore, that astronomical accuracy must at some point give way to convenience . . . , the Church has drawn up a lunar calendar which maintains as close a relation with the astronomical moons as is practicable. . . ."

Quite simply, the Church uses an idealized, formulaic full moon date rather than a true astronomical full moon to determine when Easter falls. Likewise, the idealized vernal equinox, March 21, is used instead of the true equinox, which may or may not occur on that date. The first of these formulaic full moons occurring on or after March 21 is known as the *paschal full moon*, and Easter is properly defined as the Sunday immediately *after* the paschal full moon (PFM). Thus, determining the date of the PFM in a given year is at the heart of the Easter algorithm. In order to make use of his original symbols, it will be convenient to introduce Gauss's algorithm before we address the PFM problem.

Gauss's Easter Algorithm

Here is Gauss's Easter algorithm in his own words and symbols [4]:

Complete General Rules for the Calculation of the Easter Date for the Julian, as Well as the Gregorian Calendar

If the result of the division of	by	is the remainder
the year number	19	a
the year number	4	b
the year number	7	c
the number $19a + M$	30	d
the number $2b + 4c + 6d + N$	7	e

Then Easter falls on the $(22 + d + e)$th of March or the $(d + e - 9)$th of April.

The symbols M and N are described (in Gauss's own words) as follows:

> M and N are numbers that have unchanging values for all time in the Julian calendar, and always throughout at least 100 years in the Gregorian calendar; in the former, $M = 15$ and $N = 6 \ldots$
>
> In general, one can find the values for M and N in the Gregorian calendar for any given century from $100k$ to $100k + 99$ through the following rule:
>
> Suppose that k divided by $\{^3_4\}$ gives the (entire) quotients $\{^p_q\}$ where no consideration is given to the remainders. Then $\{^M_N\}$ is the remainder one obtains, when one divides $\{^{15 + k - p - q}_{4 + k - q}\}$ by $\{^{30}_{7}\}$.

With these descriptions of a, b, c, d, e, k, p, q, M, and N, Gauss's algorithm is complete, but far from easily understood. We shall now proceed with our examination of the formulas for these 10 values that lie at the heart of the algorithm.

The Metonic Cycle and the PFM Date

The Julian calendar, in use until the late sixteenth century (and still later in England and its colonies), consists of ordinary years of 365 days and leap years of 366 days, achieved by the familiar rule of inserting February 29 into years divisible by four. Thus, the average Julian year is exactly 365.25 days. (The Gregorian calendar, to be discussed later, modifies this plan slightly.)

By the fourth century, it was known that 19 (average) Julian years span almost exactly the same length of time as 235 lunar months (new moon to new moon). This equivalence gives an average lunar month of $19 \times 365.25/235 = 29.53085$ days, whereas the true lunar month is approximately 29.53059 days. This close agreement forms the basis of the so-called *Metonic cycle,* a tabulation of new moons formulated roughly as follows. Starting with an observed new moon December 24, 322 AD, new moons are inserted into the table in intervals alternating between 30 and 29 days: January 23, February 21, March 23, . . . , December 13 in the first year, then January 12, February 10, . . . in the second year, etc. This pattern will, of course, yield an average lunar month of 29.5 days, which is slightly too short. Thus, six "leap months" of 30 days and one of 29 days were inserted in the 19-year cycle at intervals selected to keep the Metonic new moons as close to the true new moons as possible. The Metonic cycle asserts that this pattern repeats *exactly* every 19 years. More information about the Metonic cycle can be found in [2] and [7].

Though the construction of the entire table of Metonic new moons is a bit of a puzzle, only those corresponding to the first full moon on or after March 21 are critical to the Easter date. Over the entire 19-year cycle, these new moons occur (with M for March and A for April) M23, M12, M31, M20, M9, M28, M17, A5 (=M36), M25, M14, A2 (=M33), M22, M11, M30, M19, M8, M27, M16, A4 (=M35). Using Gauss's symbol A for the year and the familiar *mod* notation for the remainder in integer division, we see that his first computation $a = A \bmod 19$ simply determines where a given year falls in the ever-repeating 19-year Metonic cycle, and thus which of the 19 new moon dates listed above is relevant. One might also observe that each new moon date in the list can be obtained from the preceding one by subtracting 11 days or adding 19 days, operations that are closely related in modulo 30 arithmetic. In fact, it is not difficult to verify that this entire list of March and April new moon dates can be generated as March 8 + d, where $d = (19a + M) \bmod 30$, $M = 15$, and $a = 0, 1, 2, \ldots, 18$. Of course, the nth day of March with $n > 31$ must be interpreted as the $(n - 31)$st day of April; for example, when $a = 7$, we get $d = 28$, and the new moon of "March 36" actually falls on April 5. We will assume this convention for the remainder of the article.

Finally, we observe that the full moon date is always taken as the fourteenth day of the lunar cycle; that is, 13 days after the new moon. Thus, *the paschal full moon (PFM) date in year* A *is March* 21 + d in Gauss's formulation.

The Sunday Formula

In the list of Metonic cycle new moon dates in the previous section, the earliest and latest are March 8 and April 5, respectively, with corresponding full moons March 21 and April 18. Easter falls at least one day, but at most seven days after the PFM date, making March 22 the earliest and April 25 the latest possible Easter dates. Gauss sets the Easter date as March 22 + d + e, where $e \in \{0, 1, 2, 3, 4, 5, 6\}$ is chosen so that Easter falls on a Sunday (the first Sunday after the PFM). Finding e is a straightforward problem that amounts to counting elapsed days from a particular, known Sunday to March 22 + d in the given year, then determining what e must be added to make the count divisible by 7. Curiously enough, though this is perhaps the most easily understood part of the whole process, it is the part that Gauss explains most thoroughly. His plan for finding e is based on the Gregorian calendar, which we shall address later; for now, we present Gauss's plan with minor modifications to complete our work for the Julian calendar.

The number of days from Sunday, March 20, 1580, to March 22 + d + e in year A is $2 + d + e + i + 365(A - 1580)$, where i counts February 29s occurring in this interval. (Here $A \le 1582$; in years before 1580, our count of elapsed days will be negative.) Letting $b = A \bmod 4$, it is not hard to see that $i = \frac{1}{4}(A - b - 1580)$. Thus, we want to choose e to make

$$2 + d + e + \tfrac{1}{4}(A - b - 1580) + 365(A - 1580)$$

divisible by 7. Adding $\frac{7}{4}(A - b - 1580)$ will not affect divisibility by 7 and produces the simpler form

$$2 + d + e + 367(A - 1580) - 2b \tag{1}$$

With simple reductions and the introduction of the symbols $c = A \bmod 7$ and $N = 6$, one finds that the quantity in equation (1) is divisible by 7 exactly when

$$e = (2b + 4c + 6d + N) \bmod 7$$

Gauss's formulation of the Easter date in year A in the Julian calendar is complete: Fix $M = 15$ and $N = 6$. Set $a = A \bmod 19$ to determine the year's position in the Metonic cycle. Let $d = (19a + M) \bmod 30$ so that March 21 + d gives the PFM date for year a in the Metonic cycle. Let $b = A \bmod 4$, $c = A \bmod 7$, and $e = (2b + 4c + 6d + N) \bmod 7$ to find the next Sunday. Then Easter Sunday is March 22 + d + e or April $d + e - 9$, as appropriate.

The Gregorian Calendar Reform

Because the Julian calendar slightly overestimates the length of the year, Easter dates based on March 21 shifted further and further from the true vernal equinox. In 1582, Pope Gregory instituted corrective changes resulting in the so-called *Gregorian calendar,* which is still in widespread use today. First, to correct the drift in the equinox date that had already occurred after several centuries under the Julian calendar, the day after October 4, 1582, was declared to be October 15, 1582. Second, to prevent the same problem from arising in the future, the length of the average year was reduced by declaring that century years would be leap years only if divisible by 400. Thus, 1600 was a leap year, 1700, 1800, and 1900 were common years, 2000 was a leap year, etc.

Clearly these two corrections affect the dates upon which Sundays fall, a change that can be addressed by modifying the value of N. (This change is broadly described in the literature as the "solar equation.") Once again, our work closely mimics Gauss's presentation, adding a few details that he chose to omit.

Gauss counts the number of days from Sunday, March 21, 1700, to March 22 + d + e in year A as $1 + d + e + i + 365(A - 1700)$, where once again i counts February 29s in this interval. (Here $A \geq 1583$.) By setting $k = \lfloor A/100 \rfloor$ (the integer part of the quotient) and $q = \lfloor k/4 \rfloor$ and recalling that $b = A \bmod 4$, we have $i = \frac{1}{4}(A - b - 1700) - (k - 17) + (q - 4)$. (Here we are discarding February 29s in the century years and adding them back in for the years divisible by 400.) As before, we want to choose e to make

$$1 + d + e + \tfrac{1}{4}(A - b - 1700) - (k - 17) + (q - 4) + 365(A - 1700)$$

divisible by 7. By adding $\frac{7}{4}(A - b - 1700)$ and simplifying, it is not hard to show that the resulting expression is divisible by 7 exactly when

$$e = (2b + 4c + 6d + 4 + k - q) \bmod 7$$

Now the choice of $N = (4 + k - q) \bmod 7$ in Gauss's algorithm is clear.

The Gregorian calendar reform contained a third correction that does not influence the civil calendar, but does affect the Easter date. The reader may recall that the Metonic cycle produces a lunar month of 29.53085 days, compared to the true lunar month averaging 29.53059 days. Just as the incorrect length of the Julian year slowly accumulated to a noticeable error, the same was true for this incorrect length of the lunar month. The accumulated error is approximately 1 day in 312.5 years, or very nearly 8 days every 2,500 years. Thus, the Gregorian calendar reform included a one-day correction in the PFM date every 300 years, seven consecutive times, followed by a one-day correction after another 400 years, repeated indefinitely. The so-called "lunar equation" affects the PFM date by changing M (and thus d) according to this plan.

The solar equation, which resulted from removing February 29s from the calendar, affects the PFM dates just as it affects the Sunday calculation. Thus, the value of M must change by $k - q$, as did the value of N. All that remains is to incorporate the lunar equation into M, but here, unfortunately, the extraordinarily dependable Dr. Gauss has a minor failure. For the lunar equation, Gauss defines $p = \lfloor k/3 \rfloor$, then builds

$$M = (15 + k - p - q) \bmod 30 \qquad (2)$$

His error is clear: this M properly accounts for the solar equation with $k - q$, but it applies the one-day correction dictated by the lunar equation *every* 300 years, rather than the prescribed sequence of seven consecutive 300-year intervals followed by an eighth interval of 400 years.

Only mathematicians could be troubled by an error in the Easter date that won't surface for another 2,000 years, but since we have the means to correct it, let us do so!

Gauss's Error

One of the most widely quoted sources on the subject of the Easter date is J. M. Oudin's 1940 paper "Étude sur la date de Paques" [8]. Indeed, it is Oudin's Easter algorithm that appears in the U.S. Naval Observatory's almanac cited above. And it is common for those who reproduce Oudin's algorithm to do so after enthusiastically describing Gauss's algorithm, then "pulling the rug out from under it" by mentioning the error described above. (Or simply declaring that Gauss's method is wrong without understanding the nature of the error at all.)

Oudin writes, "Gauss, having forgotten to take into account these delays of the lunar equation in his formula for M, the latter . . . works only until 4199 inclusively and is found thus devoid of the character of generality that its author thought to have given it." Oudin is correct, but he writes without benefit of the ability to conduct a simple Internet search. With this modern tool, it is not difficult to find the following piece missing from his puzzle.

In the January/February 1816 edition of the *Zeitschrift fur Astronomie und verwandte Wissenschaften* (*Journal for Astronomy and Related Sciences*), we find a contribution from Gauss [5]. It is titled "Berichtigung zu dem Aufsatze: Berechnung des Osterfestes (Correction to the Essay: Computation of the Easter Date)" and gives publication information for the original August 1800 paper. Gauss writes

> . . . my establishment of that rule [for p] took no notice of the circumstance that the lunar equation, so-called by the originators of the Gregorian calendar, which is reasonable every 300 years . . . must actually become reasonable once every $312\frac{1}{2}$ years. Without my engaging the question here of whether this achieves the intended purpose, I only remark that my rule, with the arrangement of the Gregorian Calendar as it is, can easily be brought into complete agreement when one does not accept for the number p the quotient in the division of the number k by 3, as stipulated in

the citation above, rather one accepts the quotient in the division of the number 13 + 8k by 25.

And of course, when Gauss's corrected value $p = \left\lfloor \frac{13+8k}{25} \right\rfloor$ is incorporated into equation (2) for M, it exactly produces the necessary one-day changes every 300 years, seven consecutive times, followed by a one-day change after 400 years. Oudin's criticism was 124 years too late!

We close this section by noting that in the Gauss *Werke*, "Berechnung des Osterfestes" closes with "Handschriftliche Bemerkung" or "Handwritten remarks." Among these is the correct formula for p. Presumably, this was Gauss's handwritten remark, added to the manuscript sometime after his note of 1816.

The Exceptional Cases

Though it would appear that the Easter date algorithm is complete, there are two special cases, not at all obvious, that must be addressed. The reader will gain a much better appreciation of the challenges inherent in examining Gauss's work by considering these exceptions in his own words:

> From the above rules, one finds unique and alone in the *Gregorian calendar* the following two exceptional cases:
>
> I. If the calculation gives Easter on the 26th of April, then one *always* takes the 19th of April. (e.g., 1609, 1989).
>
> One easily sees that this case can only occur where the calculation gives $d = 29$ and $e = 6$; d can only obtain the value 29 when $11M + 11$ divided by 30 gives a remainder that is *smaller* than 19; to this end, M must have one of the following 19 values:
>
> $$0, 2, 3, 5, 6, 8, 10, 11, 13, 14, 16, 17, 19,$$
> $$21, 22, 24, 25, 27, 29.$$
>
> II. If the calculation gives $d = 28$, $e = 6$, and meets the requirement that $11M + 11$ divided by 30 gives a remainder that is smaller than 19, then Easter does not fall, as follows from the calculation, on the 25th, rather on the 18th of April. One can

easily convince oneself that this case can only occur in those centuries in which M has one of the following eight values:

$$2, 5, 10, 13, 16, 21, 24, 29.$$

With these two exceptional cases accounted for, the above rules are completely general.

When translating Gauss, it is always useful to have a list of German synonyms for "easily!"

The first exception is straightforward, though Gauss's added remarks make it less so. We noted previously that under the Julian calendar and the original Metonic cycle, Easter dates always fall between March 22 and April 25, inclusive. This rule was continued under the Gregorian calendar. If $d = 29$ and $e = 6$, however, Gauss's formula produces a PFM date of April 19 and an Easter date of April 26, in violation of the rule. The Church's Easter tables simply shifted the PFM date to one day earlier under this circumstance (from Sunday, April 19, to Saturday, April 18), effectively replacing $d = 29$ by $d = 28$. This has no effect at all *unless* $e = 6$, that is, *unless* April 19 is a Sunday; in that case, it forces Easter to occur a week earlier, on April 19.

The comments that Gauss supplies along with the first exception are puzzling until one observes that by fixing $d = 29$ in the PFM formula $d = (19a + M) \bmod 30$, one can solve to find $a = (11M + 11) \bmod 30$ and $M = (29 + 11a) \bmod 30$. The first of these equations reduces Gauss's claim of "a remainder that is *smaller* than 19" to the simple observation that a takes on the values $0, 1, 2, \ldots, 18$. Now Gauss's list of M values can be generated by successively substituting these a values into the second equation. One should also note that Gauss's choice of 1989 as an example of the first exception is erroneous; he corrects this without comment in an article on the Easter date that appeared in *Brunswick Magazine* [6] in 1807, replacing 1989 with 1981.

The second exceptional case is a result of the change described in the first. The Gregorian calendar reformers wished to preserve a basic feature of the original Metonic full moons: the PFM date is never duplicated within one 19-year cycle. But now this is sure to happen when $d = 28$ and $d = 29$ occur within the same 19-year cycle, because the $d = 29$ PFM (April 19) has been shifted to the $d = 28$ PFM date

(April 18). Gauss's peculiar way of identifying these cycles in which both $d = 28$ and $d = 29$ occur needs a closer look.

Suppose that for a given M value, there is an $a \in \{0, 1, \ldots, 18\}$ for which $d = (19a + M) \bmod 30$ produces $d = 28$; suppose also that "$11M + 11$ divided by 30 gives a remainder that is smaller than 19." The latter requirement means that there is an $a \in \{0, 1, \ldots, 18\}$ for which $a = (11M + 11) \bmod 30$. As in the first exception, this can be rearranged to obtain $29 = (19a + M) \bmod 30$; that is, for the given M value $d = 29$ is also achieved.

So how do we avoid the duplication of PFM dates when a given M value produces both $d = 28$ and $d = 29$ in a 19-year cycle? Simply shift the $d = 28$ PFM date (April 18) to that of $d = 27$ (April 17)! The apparent cascade of shifts that might be produced by this plan does not occur because the 19-year cycles containing both $d = 28$ and $d = 29$ never contain $d = 27$. (Probably the simplest way to see this is by brute force: construct a table of $d = (19a + M) \bmod 30$ values for $a = 0, 1, 2, \ldots, 18$ and $M = 0, 1, 2, \ldots, 29$.) Since $e = 6$ in Gauss's second exceptional case, the PFM associated with $d = 28$ falls on *Sunday*, April 18, so reducing d to 27 places the PFM on Saturday, April 17 and Easter on Sunday, April 18, just as Gauss's second exception dictates.

Finally, to find the eight values of M supplied by Gauss in the second exceptional case, it is probably simplest to fix $d = 28$ and consider "the requirement that $11M + 11$ divided by 30 gives a remainder that is smaller than 19" by listing the triples

$$(a,\ M,\ (11M + 11) \bmod 30)$$

for $a = 0, 1, 2, \ldots, 18$. Those meeting the requirement are (11, 29, 0), (12, 10, 1), (13, 21, 2), (14, 2, 3), (15, 13, 4), (16, 24, 5), (17, 5, 6), and (18, 16, 7), from which we can read off the middle entries as Gauss's eight M values.

Gauss's method is complete!

Conclusion

The mathematics in Gauss's Easter algorithm is trivial. But his ability to transform a desperately arcane set of rules and tables into a simple arithmetic process illustrates once again his uncanny ability to see deeply into complicated matters. If only he would give mere mortals a little more guidance through his work . . . ah, but perhaps not. For then the joy of completing the many puzzles he has left behind would be lost to the rest of us!

References

[1] Bien, R. (2004). Gauss and Beyond: The Making of Easter Algorithms, *Arch. Hist. Exact Sci.* 58: 439–452.

[2] Bond, J. J. (1875). *Handy-Book of Rules and Tables for Verifying Dates with the Christian Era, &c.* London: George Bell and Sons.

[3] Dunnington, G. W. (1955). *Carl Friedrich Gauss: Titan of Science.* New York: Hafner.

[4] Gauss, C. F. (1800). Berechnung des Osterfestes, *Werke.* 6: 73–79.

[5] Gauss, C. F. (1816). Berichtigung zu dem Aufsatze: Berechnung des Osterfestes. *Werke.* 11(1): 201.

[6] Gauss, C. F. (1807). Noch etwas über die Bestimmung des Osterfestes. *Werke.* 6: 82–86.

[7] Kennedy, T. (1909). Epact, *The Catholic Encyclopedia*, vol. 5. New York: Robert Appleton Company, available at http://www.newadvent.org/cathen/05480b.htm.

[8] Oudin, J. M. (1940). Étude sur la date de Paques. *Bulletin Astronomique* 12: 391–410.

[9] Richards, E. G. (2012). Calendars. In: P. K. Seidelman, S. E. Urban, eds. *Explanatory Supplement to the Astronomical Almanac*, 3rd ed. Mill Valley, CA: University Science Books, pp. 600–601.

Mathematical Knowledge and Reality

Paul Thagard

Since Plato, many philosophers have looked to mathematics to provide a guide for philosophy that could share in the certainty and necessity that mathematics seems to provide. This influence has been unfortunate because it provides a distorted image of how philosophy can accomplish its aims by gaining understanding of knowledge, reality, morality, meaning, and beauty. My book *Natural Philosophy* (Thagard 2019b) exhibits the advantages of taking science as more central to philosophy than mathematics. Moreover, the image of mathematics as possessing a priori certainty and necessary truths is open to challenge.

Nevertheless, no system of philosophy would be complete without an account of mathematics that explains its practice and applicability to the world. Mathematical knowledge has great practical importance, as shown in applications to technology and science, but it also seems to possess a pristine elegance that makes it less open to challenge than other fields. There is currently no good account of how mathematics can have these seemingly contrary properties of being both practically useful and amazingly solid.

Natural philosophy should be able to furnish an account of knowledge and reality in mathematics, but sadly there are serious gaps in current knowledge about how mathematics works in minds, society, and the world. I hope these gaps will be filled by future research in science and philosophy, and I will try to point in some useful directions. I propose conjectures about mathematical knowledge and reality that need to be defended by showing that they are superior to currently available alternatives:

1. Mathematical knowledge develops in human minds through the construction of neural representations for concepts and rules.

2. This knowledge results from a combination of embodied sensory-motor learning (e.g., integers) and from transbodied bindings (e.g., irrational and imaginary numbers).
3. Mathematical knowledge is not simply cognitive, for emotions also contribute to mathematical discoveries and judgments, such as appreciation of beauty.
4. Like other kinds of knowing, mathematics is not a purely individual enterprise, but it also depends on communities that provide support, inspiration, and communication.
5. Mathematical knowledge applies to the world in the same way that scientific knowledge does, as approximate correspondence to reality.

Philosophers and mathematicians have long feared that the objectivity of mathematics would be threatened by the incursion of psychological factors, but the opposite is true, just as in epistemology and ethics. Understanding how mathematics operates in human minds, including emotional and social factors in addition to cognitive ones, enhances rather than undercuts the epistemic qualities of mathematics. However, attention to the history of mathematics with a psychological perspective does challenge some exaggerated views of its certainty and necessity.

Issues and Alternatives

Mathematics raises challenging questions to be answered by competing philosophical theories. Why does mathematics seem more certain than other kinds of knowledge? Why do many philosophers think that mathematical truths are necessary, true not only of this world but of all possible worlds? Why is proof a special way of achieving knowledge? How is mathematics, which seems to transcend the world in the form of necessary truths, nevertheless applicable in practical spheres such as technology? How do finite human minds grasp mathematical ideas about infinity, for example, concerning the apparently infinite number of integers and the apparently even larger infinity of real numbers? How do mathematicians generate mathematical knowledge? What are numbers anyway? A good theory of mathematics should be able to answer these questions, or at least to show why the question is inappropriate based on false presuppositions, as I will argue for the questions about necessity and certainty.

Competing philosophies of mathematics include Platonism, logicism, formalism, fictionalism, social constructivism, empiricism, structuralism, and Aristotelian realism. *Platonism* is the view that numbers and other mathematical objects, such as sets, exist in abstract form independent of human minds and the physical world. Many mathematicians find Platonism appealing because it explains their sense that they are grasping eternal realities that do not depend on the vagaries of mind and world. Platonism would explain the apparent certainty and necessity of mathematics but has trouble with why the abstractions of mathematics are often so useful in dealing with the world of crops and computers. Platonism also seems to require dualism because only a non-material soul could somehow grasp the non-material reality of completely abstract entities. Brains are just not up to the job.

To counter the metaphysical extravagance of Platonism, many alternative philosophies of mathematics have been developed. One popular view in the 20th century was *logicism*, which claims that all of mathematics can be reduced to formal logic. The hope was that logic, with basic rules such as *modus ponens* and obvious mathematical principles such as $X = X$, could be used to generate all truths of arithmetic. This hope was dashed by Gödel's proof that any formal system that is both consistent and adequate for arithmetic is incomplete: there are statements that cannot be proved or disproved in that system.

Formalism is a philosophy of mathematics that treats mathematics as a set of marks on paper, a syntactic exercise that uses proofs to derive formulas from other formulas. This approach has problems with both the certainty of mathematics and its applicability to the real world. Similarly, the claim of *fictionalism*, that math is just make believe, has trouble explaining its usefulness and convincingness. Another skeptical view is *social constructivism*, which emphasizes the cultural aspects of mathematics.

Empiricism claims that all knowledge derives from the senses. From this perspective, mathematical truths such as $4 + 5 = 9$ and the Pythagorean theorem are just generalizations about observed reality. As a matter of empirical fact, four objects added to five objects make nine objects. This view explains the practical applicability of mathematics to the world, but has trouble dealing with the apparent certainty and necessity of mathematics, especially as it concerns abstract entities such as infinite sets.

Structuralism is a philosophy of mathematics that denies the reality of objects such as numbers but maintains that math describes abstract structures that are the basis for mathematical truths. For example, $4 + 5 = 9$ is a truth about the structure of arithmetic, not about particular numbers. Like Platonism, structuralism sees mathematics as concerned with abstract entities, and therefore has the same difficulty of figuring out how mathematics applies to the physical world. Like structuralism in the philosophy of science, structuralism in the philosophy of mathematics has the problem of figuring out how there can be structures and relations without objects that are related. In set theory, a relation is a collection of ordered objects, which makes no sense if there are no objects to order.

The current philosophy of mathematics that fits best with what is known about minds and science is James Franklin's (2014) *Aristotelian realism*, inspired by Aristotle's claim that concepts are about the world rather than abstract forms postulated by Platonic realism. According to Aristotelian realism, mathematics is the science of quantity and structure, just as physics, chemistry, and biology are sciences about the world. To explain how mathematical knowledge develops, Aristotelian realism needs to merge with a neuropsychological account of mind.

Mathematics in the Mind

The first step toward how mathematics works in the mind, the world, and society is a theory of the mental representations and processes required for the development and learning of mathematics. Platonism obviously fails as the psychology of mathematics because the abstract nature of mathematical objects and truths would divorce them from explanation based on mental mechanisms such as neural processing: there is no apparent way that neurons could interact with numbers as abstract ideas. Plato thought that all mathematical concepts, and in fact all concepts, are innate, with experience merely reminding people about abstract ideas that their souls knew before birth.

This view is problematic for many reasons, from the lack of evidence that souls exist before birth to the difficulty of understanding how experience could reactivate such knowledge. Perhaps some basic concepts about objects and quantity are innate, but not numbers like 10^{100} and relatively new areas of mathematics, such as calculus and abstract algebra. More plausibly, mathematical knowledge consists of learned concepts and beliefs that include axioms and theorems, all comprehensible as neural representations.

Mathematicians are mature humans with sophisticated linguistic capabilities, so it is tempting to think of mathematical knowledge as a matter of language. But some primitive numerical abilities are found in nonhuman animals and nonvocal human infants. For example, fish, bees, pigeons, and rats can all detect differences in small numbers of objects, as can infants only six months old. Stanislas Dehaene (2011) argues that animals and humans have the basic ability to perceive and manipulate numbers. Other theorists prefer to explain nonlinguistic numerical abilities as resulting from a more basic sense of magnitude, with size more basic than counting. Either way, we need to recognize that people's ability to do mathematics is not just symbol manipulation, but depends on more primitive sensory and motor abilities.

Such multimodal abilities are explained by viewing mental representation as semantic pointers, which are neural representations that can be formed by sensory inputs but also can be combined into new representations that go beyond the senses (Eliasmith 2013, Thagard 2019a). Semantic pointers accommodate verbal information bound into larger packages, where these packages can also include sensory-motor inputs. Groups of neurons can detect visual patterns, such as small groups of numbers, and bind these into larger representations such as *three cats* and *many dogs*. As Thagard (2019a, 2019b) argued for concepts in general, the sensory and the motor components of semantic pointers allows them to account for embodied cognition in which thoughts are strongly affected by sensory and motor representations. Chris Eliasmith's Semantic Pointer Architecture has been used to detect visual numerals and to do simple reasoning about numbers represented by convolutions of numeral shapes and relational information about number order.

Moving beyond sensory-motor inputs, can the semantic pointers explain important mathematical concepts such as *number*, *set*, *infinity*, and *shape*? Google's dictionary defines a number as "an arithmetical value, expressed by a word, symbol, or figure, representing a particular quantity and used in counting and making calculations and for showing order in a series or for identification." This definition would be more impressive if arithmetic were not circularly defined by the dictionary as the branch of mathematics dealing with numbers.

TABLE I. Analysis of the Concept *Number* Using
the Method of Three-Analysis (Thagard 2019a)

Exemplars	2, 3, 10, 100, π, etc.
Typical features	Expressed by words such as "two," represents a quantity in a collection of units, used in counting, used in calculations, ordered
Explanations	Explains: Patterns of things in the world, spatial and temporal structures, ability to count and calculate
	Explained by: Mental mechanisms of representation and manipulation

Alternatively, Table 1 presents an analysis of the concept of *number* in line with the theory that concepts are patterns of neural activation capable of integrating exemplars, typical features, and explanations. It might seem that all numbers are of equal importance, but most people are much more familiar with low numbers, such as 2 and 3, as well as powers of 10. So 149,356 is still a number, but not a standard example for most people. Mathematicians become familiar with many numbers that are mathematically interesting, such as π (3.14159 . . .), e (2.71828 . . .), and googol (10^{100}).

What counts as a number has expanded over the history of mathematics. Some cultures have only a limited range of number concepts, such as *one, two,* and *many,* whereas the ancient Greeks developed ideas about the natural numbers that can grow very large. But they lacked the concept of zero as a number, which was a development in ancient India. Early mathematicians were suspicious of negative numbers, such as -2, and later extensions to imaginary numbers such as the square root of -1 and infinite sizes of infinite numbers have provoked much consternation. The ancient Greeks did not want to believe that there are irrational numbers that cannot be expressed as the ratio of integers, but proof of the irrationality of the square root of 2 showed that the concept of number has to expand beyond familiar exemplars.

The typical features such as counting apply to the most familiar numbers but become more problematic when extended to more abstract quantities, such as π and e. Quantity does not serve well as a rigorously defining feature of number, since quantity is often defined in terms of number. Concepts of counting and quantity have important sensory-motor aspects because people often count by pointing to one thing after another, using visual, tactile, and motor representations. Then the typical features of numbers are not purely verbal, but connected to sensory-motor representations that can be bound into concepts of number used for more abstract calculations, such as using π to figure out the area of a circle. The ordering aspect of numbers is a typical feature that is embodied when numbers are viewed as arranged on a line.

The concept *number* makes explanatory contributions, for example, in tracking changes in the world, such as how two rabbits can quickly generate large numbers of rabbits. Numbers also explain practices such as being able to divide things among people, including fractional divisions, such as pieces of orange. The explanatory role of numbers becomes even more impressive when it is connected with mathematical principles in physics and other fields that deal with quantities. My attempt to explain numbers in terms of their mental uses is highly contentious, with many issues to be worked out.

Nevertheless, even in the absence of a tight definition, we can understand people as having a concept of number that combines exemplars, typical features, and explanations. This combination is explained as a neural process by the semantic pointer theory of concepts. I am not claiming that numbers are semantic pointers, just that concepts of numbers are semantic pointers. The relation between things and concepts of things is discussed in Thagard (2019b, Chap. 3) and will be further elucidated here in the section on math in the world.

The discussion of the sensory and motor aspects of quantity, counting, and order might give the impression that all mathematical concepts are embodied and learned by experience. But we have seen that the semantic pointer theory of concepts also explains how concepts can be transbodied, going beyond sensory-motor representations by recursive binding. Contrary to the empiricist philosophy of mathematics, concepts are not restricted to generalizations from observations, but they can also be constructed by combination using the process of convolution of patterns of neural firing.

Like concepts about theoretical entities discussed in Chapter 4 of Thagard (2019b), mathematical abstractions can be formed by combining existing concepts. For example, the concept of *prime number* does not have to be learned inductively from examples, although it may have some exemplars, such as 3, 5, and 7. Rather, the concept of prime number comes from combining concepts of number, divisibility, and negation. Conceptual combination carried out by the neural process of convolution quickly takes us beyond the limitations of embodiment.

Axioms and Theorems

Concepts say nothing about the world without being combined into beliefs. The number 4 alone makes no claims, but the belief that $4 + 5 = 9$ asserts that if you have four things and five things then you have nine things. Mathematical truths do not have to be just generalizations about the observable world because conceptual combination by binding into semantic pointers allows construction of beliefs that usefully transcend the senses. Arithmetic does serve to characterize quantities in the world but can also be based on different kinds of coherence. When mathematical principles are part of scientific theories, they gain explanatory coherence based on how well they jointly explain evidence. But math is often different from science in consisting of axioms and theorems rather than hypotheses and explanations.

Like arithmetic, geometry contains general statements, but they usually concern shapes rather than quantities. The Pythagorean theorem can be expressed in various ways, for example, in the words that in a right triangle the square of the hypotenuse is equal to the sum of the square of the other two sides. Alternatively, the theorem can be represented jointly by the diagram in Figure 1, which shows the lengths of the respective sides, and by the equation: $a^2 + b^2 = c^2$. This theorem requires several concepts, including *triangle, right angle,* and *hypotenuse,* but combines them into a new claim supported both by observations in the world and by proof from axioms. The words, equation, and image are complementary, and all can be formed into mental representations as semantic pointers using the techniques spelled out in Thagard (2019a).

Because of the historical influence of Euclidean geometry and later views, such as logicism, mathematical knowledge is often construed as a system in which theorems are derived from a small set of axioms and definitions. Ideally, the axioms should be self-evident and the methods of proof for establishing theorems should be deductively

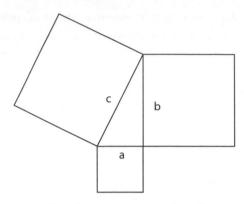

FIGURE 1. Visual representation of a right triangle.

precise, resulting in a whole body of knowledge that is unassailable. The section of this article on coherence suggests that this unassailability is exaggerated.

Nevertheless, philosophy of mathematics has to take seriously the role of axiomatic systems in mathematical practice. How can we understand axioms, theorems, and deductive inference within the Semantic Pointer Architecture? Axioms are easily understood as multimodal rules with an if-then structure, where the *if* and *then* parts can be verbal or sensory-motor. For example, Euclid's first postulate, that a straight line can be drawn between any two points, becomes the verbal rule: if you have two points, then you can draw a line between them.

But mental representation of this postulate is not simply a matter of binding linguistic entities into larger linguistic entities because the semantic pointers that are bound together into the larger postulate can include sensory-motor aspects. For example, *draw between* is a motor activity, as well as one that is visually perceived. Both the points and the line connecting them are also naturally encoded visually. So geometrical axioms and the theorems that depend on them are multimodal, as naturally captured by multimodal rules consisting of semantic pointers.

Outside of geometry, axioms and theorems are more likely to be just verbal, for example, the first two axioms of Peano arithmetic: 0 is a natural number, and for any natural number x, $x = x$. The crucial link between axioms and theorems is proof, a series of logically correct steps from the axioms and definitions to the theorem. Logical steps include ones performed by *modus ponens*: P and *if* P *then* Q, therefore

Q. Eliasmith (2013) shows that the Semantic Pointer Architecture can carry out *modus ponens*, but simulations of other aspects of mathematical proof are daunting challenges for future work.

METAPHORS AND ANALOGIES

Contrary to the impression given by axiomatic systems, mathematics has much substantial cognitive flexibility that can be seen in the role of metaphors, analogies, and heuristics. George Lakoff and Rafael Núñez (2000) describe how arithmetic grows out of a metaphor of object collection, based on a mapping from the domain of physical objects to the domain of numbers. The actions of collecting objects of the same size and putting collections together map onto addition, and taking a smaller collection from a larger collection maps onto subtraction. Even mathematical ideas about infinity have metaphorical connections with processes that have beginning and extended states. Lakoff and Núñez describe mathematical knowledge as developing from a combination of embodied metaphors and conceptual blending. Thagard 2019a (Chapter 10) shows how metaphor and blending are naturally explained in terms of semantic pointers, but much theorizing and simulating is required to support the hypothesis that mathematical metaphors are the result of neural mechanisms.

Complex metaphors rely on analogies that map between two domains, and many historical examples show the value of analogical thinking for developing mathematical knowledge. For example, Descartes developed a new approach to geometry by analogy to algebra. Thagard 2019a (Chapter 6) proposes a theory of analogy based on semantic pointers, but the applicability of this account to mathematics needs to be developed. Similarly, the heuristics that George Polya (1957) argued are important in mathematical thinking, such as decomposing a problem, should be translatable into multimodal rules with semantic pointers, but much research is required to carry out the translation.

COHERENCE AND NECESSITY

Why does mathematics seem so much more certain than ordinary empirical knowledge? It is easy to imagine crows that are not black, but hard to see how 3 plus 5 could be other than 8. The arguments in Thagard (2019b, Chap. 3) against a priori truths should extend to mathematics but leave the problem of explaining why mathematical truths seem so solid on grounds other than pure intuition.

My answer is that mathematical certainty is an extreme case of the high confidence that derives from multiple kinds of coherence. For example, we have three strong and interlocking reasons for thinking that the Pythagorean theorem will continue to hold. First, it matches innumerable observed cases where the size of hypotenuses has been calculated and turned out to conform to the theorem. That the theorem is actually true explains the evidence that it seems to be true.

Second, the Pythagorean theorem is deductively coherent with Euclid's axioms because of the proof that derives it from the axioms. There is no need to suppose that the axioms are self-evident because they also achieve coherence with observations and the many other plausible theorems that can be inferred from them.

Third, the Pythagorean theorem shares in the explanatory coherence of the scientific and technological fields that employ it. For example, geography uses it to measure distances, and architecture uses it to construct buildings. For all these kinds of coherence, the Pythagorean theorem is so centrally embedded in our web of belief that it has become hard to imagine that it is false.   THAGARD LOVES THAGARD

Thagard (2019, Chap. 3) argues that imagination is not a good guide to truth because it is constrained by prejudices and ignorance. In contrast, the three kinds of coherence that support mathematics are each reliable in establishing sustainable truths, and pooling them creates even more reliability. Like geometry, arithmetic also benefits from this triple coherence because it coheres with observations, axiomatic proofs, and usefulness in science and technology.

Certainty is never absolute, in keeping with the emphasis on fallibility in Thagard (2019b, Chap. 3). There have been a few historical cases where established mathematical views have been overturned, for example, Euclid's fifth postulate, which implies that at most one line can be drawn parallel to a given line through a point not on the line. Einstein's theory of relativity acquired substantial evidential support while employing non-Euclidean geometry that rejects this postulate. The apparent certainty of mathematics comes from its overall coherence, not from the metaphysical guarantee of Platonism, the purely deductive coherence of logicism, or the triviality of formalism.

My coherentist justification of mathematical knowledge cannot accommodate another apparent feature of it, necessary truth. Math is supposed to be true of all possible worlds, not just this one, as shown by our inability to imagine its falsehood. But Thagard (2019b, Chap. 3) argued that such failures are often signs of ignorance rather than profundity, so the best strategy is what Thagard (2019b, Chap. 5) called *eliminative explanation*. Necessary truths in general can be explained away as an illusion resulting from insufficient understanding of the interactions of mind and world. The illusion can be replaced by appreciation of the objectivity and less-than-absolute certainty that comes from multiple kinds of coherence.

EMOTIONS

Mathematics is supposed to be the epitome of cold reason, but emotions are important for mathematical cognition in several ways. First, according to G. H. Hardy (1967), a mathematician is a maker of patterns whose importance is judged in part by their beauty.

> The mathematician's patterns, like the painter's or the poet's must be *beautiful*; the ideas like the colours or the words, must fit together in a harmonious way. Beauty is the first test: there is no permanent place in the world for ugly mathematics.

The experience of ugliness is an unpleasant emotion to be avoided as much as the positive experience of beauty is to be pursued. Thagard (2019b, Chap. 9) shows how beauty is an emotional response to the coherent unification of attractive diversity, and Hardy is right that mathematical beauty is the same. Elegance results from unification of diverse theorems by basic axioms and efficient proofs.

Second, in order to pursue the thousands of hours it takes to become an accomplished mathematician, people have to find it at least interesting, an epistemic emotion accompanied by moderate physiological arousal and modest goal relevance. Third, in order to put in the much greater degrees of effort required to make substantial breakthroughs, mathematicians need to find their work exciting, a much stronger epistemic emotion. Fourth, emotions also operate in the social context of mathematics, for example, when mathematicians evaluate each other's work as being surprising rather than dull, or as an exciting breakthrough rather than just a mundane extension.

Mathematics in the World

Some philosophers have argued that we should believe in numbers and other mathematical structures, not for Platonist reasons, but simply because they are indispensable for doing science. But what does indispensability amount to? Mathematical areas such as arithmetic and geometry are useful for generating explanations and producing calculations that are part of technology and science. Thagard (2019b, Chap. 4) argues that technology contributes to inferring that science is about the world, and a similar argument works for math.

If the Semantic Pointer Architecture shows how mathematics can operate in minds and brains, why is math so amazingly useful in the world? Platonism faces the problem of figuring out why heavenly ideas are so useful in worldly applications ranging from building bridges to quantum computing. One solution to this problem would be to meld Platonism with idealism. If there is no world external to human minds, and if mathematical ideas are just abstractions in human minds, then the mind-to-world connection is unproblematic because everything is mind. But Thagard (2019b, Chap. 4) argues that not everything is mind, which is a relatively recent addition to the universe, according to available evidence. So the math-to-world connection remains problematic.

One straightforward solution to this problem would be the empiricist one that mathematical statements are just generalizations about the world, so it is automatic that they apply to the world. However, like the exaggeration that all mathematics is embodied, empiricism is incompatible with recognition that mathematics operates with many layers of abstraction such as sets, groups, and geometric manifolds that go far beyond ideas that fit with sensory and motor experience.

Formalism and logicism also have serious problems with applicability. If mathematics is just a system of marks or consequences of logical truths, its applicability to the world remains unexplained. Fictionalism says that mathematical objects are fictions, just like unicorns, but is flummoxed about why arithmetic, geometry, and other branches of mathematics are so much more useful for dealing with the world than theories of unicorns. *NO THAGARD CITATION?*

Another misguided way of understanding the relation between mathematics and the world is to say that the world is fundamentally mathematical. This view originated with the followers of Pythagoras, who were so impressed by the applicability of mathematics to the world that they concluded that the world is nothing but mathematics, with numbers as the fundamental reality. The Pythagorean fallacy is to leap from the historical observation that mathematics has been enormously valuable in describing and explaining the world to the conclusion that reality is just mathematics.

Max Tegmark (2014) uses the following argument in his book *Our Mathematical Universe*.

1. There exists an external reality completely independent of humans.
2. Physics aims toward a complete description of this external reality in the form of a Theory of Everything.
3. The Theory of Everything would avoid words and concepts in favor of mathematical symbols.
4. Therefore, our external physical reality is a mathematical structure.

I agree with step 1, but the rest of the steps are shaky.

Even if physics manages to come up with a theory that unites general relativity and quantum theory with a new account of quantum gravity, the result would not be a Theory of Everything. For reasons given in the discussion of reduction and emergence in Thagard (2019b, Chap. 5), it is implausible that quantum gravity would tell us much about biochemistry, evolution, neuroscience, psychology, and sociology. Moreover, we currently have no idea what the theory of quantum gravity will look like or if it will ever be produced. The current candidates of string theory and loop quantum gravity use lots of math, but they also abound with concepts that help to make the symbolic equations meaningful. If quantum gravity was just math with uninterpreted symbols, people might be able to use it to make predictions, but they would still find it unsatisfying in its failure to provide explanations. The legitimate aims of science include understanding, not just prediction. That is why there are now more than a dozen competing interpretations of quantum theory from scientists and philosophers not content with its impressive predictive accomplishments.

The philosophy of mathematics needs a metaphysical theory that complements the epistemology based on semantic pointers and coherence. The strongest current candidate is Aristotelian realism, which is realist in that it takes mathematical claims to be true or false about the world. According to Franklin (2014), there were mathematical properties and relations in the world long before people came along to develop mathematical knowledge. There were quantities of stars and other objects, and many other properties in relations, such as symmetry and ratio. Mathematics is the science of quantities and structures in the world.

The main problem with this view is that much of mathematics does not seem to be about the world. For example, transfinite set theory developed from Cantor's creative proofs that there is an infinite number of infinite sets of different sizes. Many areas of mathematics developed in isolation from physical considerations of the world, for example, group theory and the theory of complex numbers, which only later turned out to be useful in physical applications, such as Schrödinger's equation in quantum theory.

I think that the applicability of mathematics to the world is best explained by two factors. First, much of mathematics is motivated by the desire to understand the world, from descriptive generalizations to explanatory ventures in physics. Hence, many mathematical concepts, such as those in basic arithmetic and geometry, are at least partly embodied and tied to empirical generalizations, giving them word-to-world meaning.

Second, conceptual combination allows mathematical concepts to transcend the world, with merely word-to-word meanings akin to those found in fictional novels, plays, and films. But just as such works sometimes turn out to reveal a lot about the world, for example, in the psychological insights of Leo Tolstoy and Jane Austen, some of the transcendent concepts that mathematicians develop turn out to be useful for physicists and other scientists. Hence, latterly valuable concepts like complex numbers and non-Euclidean geometry begin to acquire word-to-world semantics as well, just as Aristotelian realism requires.

My metaphysics of mathematics is thus a combination of realism and fictionalism, varying with different domains. For mathematical domains derived from sensory-motor experience or connected with scientific theories, realism rules. But for domains that merely reflect the creative power of conceptual combination, the products are best viewed as fictions, although they may eventually turn out to be connected with the quantities and structure in the physical world when they are incorporated into scientific theories.

So what are numbers? The psychological suggestion that numbers are concepts understood as patterns of neural firing is not plausible because there seem to be far more numbers, perhaps infinitely more, than there are brain-bound concepts limited by measly billions of neurons. For now, we can only say how the concept of number functions, as neural patterns with the typical features laid out in the analysis in Table 1. I am not happy with any of the standard views that numbers are abstract entities or fictions, but a better account remains to be developed.

Mathematics in Society

Postmodernists deny both mind and world and see reality and mathematics as socially constructed. I have argued that math is tied to mind *and* world, but there are undoubtedly ways in which mathematics is importantly social. Historically, the problems that mathematicians work on have often been shaped by social circumstances with respect to enterprises such as agriculture, finance, business, technology, war, and science in general. Recognizing that mathematicians are affected by their social contexts is very different from saying that these contexts determine the content of their mathematics—the concepts, axioms, proofs, and theorems that they develop.

There have been social influences on mathematical controversies, for example, in the battle between the continental approach to the calculus developed by Leibniz versus British adherence to Newton's notation. But such controversies are resolved by mathematical criteria rather than social power.

Nevertheless, the practice of mathematics has identifiable social mechanisms. The proofs proposed by one mathematician have to be checked by other mathematicians, who sometimes discover errors to be corrected. It takes much collective work to verify and improve complex proofs, such as Andrew Wiles's proof of Fermat's Last Theorem. Like scientists, mathematicians sometimes collaborate to produce work that they would not be able to accomplish alone. An extreme example is Paul Erdős, who had hundreds of collaborators. Mathematicians, like scientists and artists, often belong to emotional communities on which they depend for support and inspiration. Hence, mathematics depends on social mechanisms of verbal and non-verbal communication.

I have tried to show how Franklin's Aristotelian realism can be enriched by Eliasmith's Semantic Pointer Architecture to provide a new approach to the philosophy of mathematics. To make this venture convincing, my speculations about mathematical concepts and theorems as semantic pointers have to be tested by the standard cognitive-science method of building computational models that simulate the thinking of mathematicians and ordinary people doing math. I hope that future philosophers and theoretical neuroscientists will pursue these tasks. The result would be a naturally superior alternative to the rationalism and empiricism that have dominated the philosophy of mathematics.

A FLAWED, EGOTISTIC, NAIVE ARTICLE

Acknowledgments

Adapted from Thagard (2019b), pp. 272-286.

References

Dehaene, S. (2011). *The Number Sense: How the Mind Creates Mathematics,* 2nd ed. Oxford, U.K.: Oxford University Press.

Eliasmith, C. (2013). *How to Build a Brain: A Neural Architecture for Biological Cognition.* Oxford, U.K.: Oxford University Press.

Franklin, J. (2014). *An Aristotelian Realist Philosophy of Mathematics.* London: Palgrave Macmillan.

Hardy, G. H. (1967). *A Mathematician's Apology.* Cambridge, U.K.: Cambridge University Press.

Lakoff, G., and Núñez, R. E. (2000). *Where Mathematics Comes From: How the Embodied Mind Brings Mathematics into Being.* New York: Basic Books.

Polya, G. (1957). *How to Solve It.* Princeton, NJ: Princeton University Press.

Tegmark, M. (2014). *Our Mathematical Universe.* New York: Knopf.

Thagard, P. (2019a). *Brain-Mind: From Neurons to Consciousness and Creativity.* New York: Oxford University Press.

Thagard, P. (2019b). *Natural Philosophy: From Social Brains to Knowledge, Reality, Morality, and Beauty.* New York: Oxford University Press.

The Ins and Outs of
Mathematical Explanation

Mark Colyvan

Gauss referred to mathematics as the queen of sciences. His remark was intended to flag the privileged and exalted position that mathematics occupies within the sciences. And just as royalty is set apart from its subjects, so too is mathematics set apart from the rest of science. Mathematics is certain, its results stand for all time, and it proceeds by a priori methods. The rest of science, by contrast, is uncertain, fallible, and a posteriori. In what follows, I'll highlight another interesting difference between mathematics and the rest of science. This difference concerns the way explanation operates in mathematics and the surprising way mathematical explanations can feature in broader scientific explanations.

First I provide a bit of background on scientific explanation. There are a number of important questions that naturally arise about explanation. When is an explanation required? How do we recognize explanations as explanations? What is an explanation? There are no easy answers to any of these questions, but for present purposes we can take partial answers to the first two questions to be that an explanation is called for when an appropriate "why" question is pressing. Why did the car skid off the road? Why did the anode emit x-rays? An explanation in its most general sense is an answer to such "why" questions, and in answering such questions, explanations reduce mystery. Exactly how an explanation reduces mystery brings us to the third question.

There are many competing philosophical accounts of explanation, and I won't try to do justice to them here. The topic has a long and distinguished history going back to ancient times. In what follows, I will explore two types of mathematical explanation and demonstrate how they put pressure on one common and intuitively appealing account of explanation. The account in question, the causal account, holds that an explanation consists of providing the causal history of the event in need of explanation. After the causal history is presented, the event in question is no longer thought to be mysterious (e.g., because the posterior probability of the event, conditional on the causal history, is higher than the unconditional prior probability of the event in question). Now I turn to mathematical explanation.

Intramathematical Explanation

It is well known that some, but not all, proofs are explanatory. Explanatory proofs tell us *why* the theorem in question is true, whereas the nonexplanatory proofs merely tell us *that* the theorem in question is true [11]. With an explanatory proof, we have one mathematical fact being explained by another (or other) mathematical fact(s). Call such cases of explanation within mathematics *intramathematical explanations*.

An example will help. Consider the well-known Euler result that there is no way of traversing the seven bridges of Königsberg once and only once in a single trip, beginning and ending in the same place.

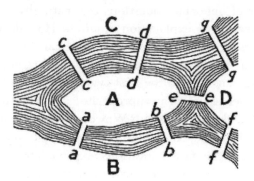

FIGURE 1. The seven bridges over the River Pregel in Königsberg (in Euler's time) [18].

In its mathematical formulation, this becomes: there is no Eulerian cycle for the following multigraph.

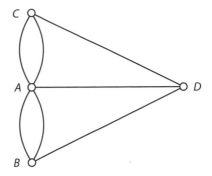

The crucial result here is that a connected graph has an Eulerian cycle iff it has no vertex with an odd degree [17]. That is, there are no vertices (or nodes) with an odd number of edges meeting there.[1] The central idea of the graph-theoretic proof is that on any Eulerian cycle, every arrival at a vertex must be accompanied by a departure; an odd valence for a vertex signifies an arrival without a corresponding departure or a departure without a corresponding arrival. Indeed, this is the core explanatory insight of the graph-theoretic proof.

Compare such a proof with a brute-force, combinatorial proof that there is no Eulerian cycle for the previously mentioned multigraph. The latter would indeed deliver the result, but armed only with such a proof, we would be none the wiser as to why there is no Eulerian cycle for the multigraph in question. We'd just know that all options had been tried and none of them worked. Moreover, the graph-theoretic proof is more general. It shows, for example, that knocking out one of the edges from C to A will not help. The brute-force method would need to start over again to show that there is no Eulerian cycle for the modified graph.[2]

What is the sense of explanation at work here? It is not the causal sense I described earlier, common in science and in everyday life. When we ask, for example, why the car accident occurred, we are typically looking for responses that appeal to driver fatigue, the icy road, the velocity of the vehicle, and so on. Sometimes, however, we might appeal to more general background conditions such as "it is a dangerous road," or "he's not a very good driver." These more general conditions offer interesting complements to causal explanations and, in some contexts, it is the more general conditions that we are interested in [3]. Still, in many legal, scientific, and everyday contexts when we seek an explanation, what we are after is the causal history of the accident: all the temporally prior events or conditions that ultimately resulted in the accident. But mathematical explanation is not in the business of providing causal histories. Talk of causes is completely out of place in mathematics. The existence of the number i *caused* the polynomial $X^2 + 1$ to have two roots? No! The existence of i is the reason that $X^2 + 1$ has two roots, but it is a category mistake to think of i as causing anything. Numbers and other mathematical objects do not seem to be the right kind of things to be in the causal nexus of the universe.[3]

It might be tempting to suggest that explanatory proofs don't provide causal explanations but something stronger: deductive explanations. After all, the conclusion of a chain of deductive reasoning, as we have in the proof of a mathematical theorem, follows of necessity. It doesn't merely follow because of contingent events in the past. Although this is right—mathematical consequence is typically deductive and this is stronger than causal consequence—deduction cannot be the key to mathematical explanations. If it were, all proofs would be explanatory, but this is clearly not the case. For example, the brute-force, combinatorial proof mentioned earlier also delivers a deductive result. But, as we have seen, this proof is not explanatory. The source of mathematical explanation must lie elsewhere.

Perhaps mathematical explanation relates to the structure of the proof.[4] We have a variety of different proof structures in mathematics: conditional proof, reductio ad absurdum, finite induction, transfinite induction, disjunctive syllogism, universal generalization, and proof by cases, to name a few. But there are problems with the view that the explanatoriness of a proof rests entirely on its structure. Apart from anything else, it is often relatively easy to transform the proof structure without any substantial change to the main body of the proof. Think, for example, of how, with a small change in the setup, a reductio proof of the infinitude of the primes can be turned into a proof by universal generalization [8].

A more promising strategy is to look below the level of the structure of the proof and at the details of the proof. Here we find two quite distinct lines of thought arising. One is that a proof is explanatory because it proceeds via the "right kind" of paths. What are the right kinds of paths? Perhaps they are paths that connect results to other results in the same domain of mathematics. This might be why elementary proofs are valued in number theory: they deliver number-theoretic results via number-theoretic means, without excursions into complex analysis. On the other hand, sometimes a proof is seen to be explanatory because it builds bridges between different areas of mathematics, often by showing that a result in one area is a special case of a more general result. Category theory proofs of results in group theory (such as the Free Group Theorem) may be instances of this [15, p. 123]. If this is right, there are two quite distinct kinds of explanation operating in mathematics: one local and the other more global and unifying.

Extramathematical Explanation

So far, we have been considering the explanation of one mathematical fact by another. Contrast this intramathematical explanation with another kind of explanation in which mathematics can feature: the explanation of some physical phenomenon by appeal to mathematics. Call the latter *extramathematical explanation*. This is more controversial. Although our understanding of intramathematical explanation leaves a lot to be desired, its existence is widely accepted. But there is a rather heated philosophical debate surrounding the very existence of extramathematical explanation.[5]

What's all the fuss about? The main reason that some philosophers object to the extramathematical explanation is that it would seem to commit them to Platonism: the view that mathematical objects, such as sets, numbers, functions, and the like, exist. Why? If you subscribe to a principle known as *inference to the best explanation*, you're committed to all the entities that play an explanatory role in your best science: electrons, quarks, black holes, gravitational waves, and so on. Add mathematical objects to this list, and it just becomes too weird for some philosophers' tastes. Unless they are willing to give up on the principle of inference to the best explanation, those who find Platonism unacceptable are forced to deny that there are any extramathematical explanations. Be that as it may, there seems to be no shortage of examples of mathematics apparently playing crucial explanatory roles in science. Here I'll quickly look at a couple of these.

The first involves the Honeycomb Theorem. Charles Darwin observed the incredible regularity of hexagons in honeybee hives and conjectured that the hexagonal shape was in some sense optimal [9]. Darwin was right, but vindication took some time. It finally came in 2001 from a mathematical theorem: the Honeycomb Theorem [12]. The Honeycomb Theorem states that a hexagonal grid represents the optimal way to divide a surface into regions of equal area with the least total perimeter. Of course, the mathematics on its own doesn't explain why honeybees build hexagonal hives. We also need some biology: wax is expensive, so bees need to minimize its use (hence, minimizing total perimeter) while maximizing the area for honey storage (there's no wasted space in the hive). It's also worth noting that this is a tiling problem, not a sphere-packing problem. This is because bees need access to the cells. Now throw in some evolutionary theory to explain why less efficient bees have been selected against, and we are left with hexagonal hives. Although the full explanation involves a mixture of biology and mathematics, arguably it's the mathematics that's doing the heavy lifting [14].

Another phenomenon whose explanation involves mathematics appears in the life cycle of cicadas. There are seven species of North American cicadas in the genus *Magicicada*, each with prime number life cycles. These cicadas emerge from the ground en masse once every 13 or 17 years, depending on the species. The explanation biologists offer for these unusual life cycles is in terms of evolutionary advantage. A periodic organism trying to avoid a periodic predator needs to minimize the number of years of overlap between itself and the predator. It is rather straightforward to prove that having a prime life cycle does this [2]. If a life cycle has a prime period, predators need very specific periods to overlap on a regular basis. For example, the predator would need the same prime life cycle, suitably coordinated. Again, we see that the explanation relies on mathematics—in this case, some elementary number theory. Moreover, the mathematics seems to be doing the important work in the explanation.

This last case also highlights an interesting general point about extra-mathematical explanations. These explanations typically tell us that not only is the world thus and so, but, in a very important sense, it *had to be thus and so*. The cicada life cycles are apparently squeezed from above by (not well-understood) biological considerations and from below by the fact that small primes are not as effective for predator avoidance. The remaining window, arguably, has four primes: 7, 11, 13, and 17. The periods found in nature are the largest primes in this set. These represent the optimal predator-avoidance strategies, subject to various biological constraints.

There are many other cases of such extramathematical explanations discussed in the literature.[6] What is common to them all is that the mathematics is crucial to the success of the explanation, and if there is a mathematics-free alternative available, such explanations are impoverished in various ways. In particular, the mathematics-free, causal explanations get bogged down in the contingent detail and fail to reveal the big picture. For example, a causal account of the Kirkwood gaps in the asteroid belt would give the causal histories of all the asteroids in the vicinity and would show why each asteroid fails to be orbiting in one of the Kirkwood gaps. Without the mathematical analysis, it looks as though it's a mere coincidence that there are no asteroids in the Kirkwood gaps. But it is no coincidence. The Kirkwood gaps are unstable orbits, and this crucial piece of the story is delivered by the mathematical explanation [8, chap. 5]. Part of the power of mathematics is that it enables abstraction away from the often irrelevant contingent details, and it goes straight to the core of the explanation.

What Is Explanation?

It appears that mathematics can deliver what we intuitively recognize as both intra- and extramathematical explanations, and, as we have seen, neither of these conform to the causal-history model of explanation. This suggests that further investigation of mathematical explanation could profitably contribute to a more general understanding of the nature of explanation. Mathematical explanation is also important for debates over Platonism and its rivals. After all, if an entity plays an indispensable explanatory role in one of our best scientific theories, this is taken by many scientists and philosophers alike to be a sure sign that the entity in question exists. The existence of extramathematical explanation would thus seem to lend support to Platonism.

Philosophers of mathematics have been largely interested in extramathematical explanation, and mathematicians have long been interested in intramathematical explanation. But the ease with which we accept the application of intramathematical explanations to physical phenomena, as in the Honeycomb Theorem case, suggests that these two kinds of mathematical explanations are closely related. Philosophers of mathematics and mathematicians really ought to hang out together more (and they would both benefit from discussions with philosophers of science). For a start, it would be extremely useful for philosophers to have a good stock of proofs considered by mathematicians to be explanatory and proofs not considered to be explanatory.[7] It would also be good to have mathematicians' thoughts on what distinguishes explanatory proofs from the others. It would be interesting to explore whether proof is the only locus of explanation. After all, if one kind of intramathematical explanation lies in unifying branches of mathematics, perhaps domain extensions and even generalized definitions could be thought to facilitate explanation. These are all issues on which this philosopher of mathematics would welcome the opinions of mathematicians.

Acknowledgments

Thanks to Clio Cresswell and Greg Restall for very helpful discussions on the topic of this article. Thanks also to Jim Brown, Mary Leng, and an anonymous *Mathematical Intelligencer* referee for detailed and constructive comments on earlier versions of the article. This work was funded by an Australian Research Council Future Fellowship (Grant Number FT110100909) and the Alexander von Humboldt Stiftung.

Notes

1. Or to revert to the original problem: The walk is possible so long as it does not involve an odd number of bridges with one end on the same land mass.

2. The graph-theoretic proof explains the *mathematical fact* that there is no Eulerian cycle for the above multigraph. That's the intramathematical explanation arising from the explanatory proof. But the proof also plays a crucial role in explaining why the previously described walk around Königsberg cannot be completed. The latter is the related extramathematical fact. Shortly, I'll have more to say about this quite different kind of explanation.

3. Indeed, this is one of the reasons many philosophers have misgivings about the existence of mathematical objects. The causal inertness of mathematical objects makes it difficult to understand how we can have mathematical knowledge [4].

4. There is also an interesting question about the status of pictures in proofs: can pictures serve as proofs, and do they, in some cases, deliver genuine understanding [5]?

5. The main proponents of extramathematical explanation are Alan Baker [2], Mary Leng [13], and me [6, 7].

6. Other examples are the Euler graph-theory explanation of why the bridges of the Königsberg walk cannot be completed, why the physical act of squaring the circle is impossible, and why the Kirkwood gaps in the asteroid belt have the specific locations they do. *See* [16] for discussions of some of these examples.

7. Some steps toward this goal have been made in [1, 10].

References

[1] Aigner, M., and Ziegler, G. M. 2010. *Proofs from the Book*, 4th ed. Heidelberg, Germany: Springer.

[2] Baker, A. 2005. "Are There Genuine Mathematical Explanations of Physical Phenomena?" *Mind*, 114(454): 223–238.

[3] Baron, S., and Colyvan, M. 2016. "Time Enough for Explanation," *The Journal of Philosophy*, 113(2): 61–88.

[4] Benacerraf, P. 1973. "Mathematical Truth," *The Journal of Philosophy*, 70(19): 661–679.

[5] Brown, J. R. 2008. *The Philosophy of Mathematics: A Contemporary Introduction to the World of Proofs and Pictures*, 2nd ed. London: Routledge.

[6] Colyvan, M. 2001. *The Indispensability of Mathematics*, New York: Oxford University Press.

[7] Colyvan, M. 2010. "There Is No Easy Road to Nominalism," *Mind*, 119(474): 285–306.

[8] Colyvan, M. 2012. *An Introduction to the Philosophy of Mathematics*, Cambridge, U.K.: Cambridge University Press.

[9] Darwin, C. 2006. *On the Origin of Species by Means of Natural Selection*. Mineola, NY: Dover. (First Published in 1859.)

[10] Davis, P. J., and Hersh, R. 1981. *The Mathematical Experience*, Boston: Birkhäuser.

[11] Gowers, T., and Neilson, M. 2009. "Massively Collaborative Mathematics," *Nature*, 461 (October 15), 879–881.

[12] Hales, T. C. 2001. "The Honeycomb Conjecture," *Discrete Computational Geometry*, 25: 1–22.

[13] Leng, M. 2009. *Mathematics and Reality*. Oxford, U.K.: Oxford University Press.

[14] Lyon, A., and Colyvan, M. 2008. "The Explanatory Power of Phase Spaces," *Philosophia Mathematica*, 16(2): 227–243.

[15] Mac Lane, S. 1998. *Categories for the Working Mathematician*, 2nd ed. New York: Springer.

[16] Mancosu, P. 2008. "Explanation in Mathematics," in E. N. Zalta (ed.), *The Stanford Encyclopedia of Philosophy* (Fall 2008 Edition), plato.stanford.edu/archives/fall2008/entries/mathematics-explanation/.

[17] Trudeau, R. J. 1993. *Introduction to Graph Theory*. New York: Dover.

[18] Weisstein, E. W. 2016. 'Königsberg Bridge Problem," *Math-World—A Wolfram Web Resource*. http://mathworld.wolfram.com/KoenigsbergBridgeProblem.html (accessed August 2016).

Statistical Intervals,
Not Statistical Significance

Gerald J. Hahn, Necip Doganaksoy, and William Q. Meeker

There is now general agreement within the statistical community that changes are needed in the way that we approach inference.[1] In particular, in many applications, the current (and long-standing) use of significance/hypothesis testing and p-values less than 0.05 needs to be replaced or supplemented by more cogent data analysis tools.

However, moving from acceptance by statisticians to application in practice is a big step.

How can we speedily get the rest of the world, in general, and statistical practitioners (i.e., users of statistical methods who are generally not professional statisticians), in particular, on board? We see two elements in making this happen.

The first is to modify the introductory statistics course—often the only statistical training that practitioners receive—to reflect current thinking; such courses have typically placed heavy emphasis on the use of significance tests.

The second element—the one we consider here—is winning over engaged practitioners typically committed to significance testing. We cannot wait for these folks to retire and be replaced by more forward-thinking analysts.

We describe an approach that we have discussed previously, first in 1974 in *Chemtech*, a now-defunct *Scientific American*-type journal for chemists and chemical engineers, published by the American Chemical Society,[2] and again in November 2017 in *Quality Progress* (*QP*), the official publication of the American Society for Quality.[3]

Our approach is twofold. First, we demonstrate via a case study from industry how statistical significance differs from practical importance.

Then, we propose and illustrate the use of statistical intervals instead. We recognize that this approach is not applicable to all situations. But we do think that when applicable—as in many industrial situations—it can accelerate the path toward de-emphasis of significance testing and help practitioners focus on more meaningful interpretation of their findings.

The Inadequacy of Significance Tests

Practitioners need to be aware that statistical significance differs from practical importance in that statistical significance is highly dependent upon sample size. For a large sample, a statistically significant result is likely to be obtained even when the actual magnitude of an effect is small and of little or no practical importance. On the other hand, for a small sample, it is quite likely that insufficient evidence of a statistically significant result will be obtained—even when there is, indeed, an effect of practical importance.

Consider this example from our *QP* article. A factory making glass products wishes to replace its current furnace, but management is concerned that this change might lead to appreciable deterioration in glass tensile strength. The furnace currently produces glass specimens with a tensile strength distribution that is well represented by a lognormal distribution (i.e., the logarithm of tensile strength is normally distributed), with a median value of 35 megapascals (MPa) and a shape parameter (standard deviation of log tensile strength) of 0.25. A drop in the distribution median from 35 MPa to 34 MPa arising from the new furnace (Figure 1) would not be of practical importance. However, a median shift from 35 MPa to 31 MPa (Figure 2) would be, and is reason to reject the new furnace.

As a first illustration, assume that the new furnace reduces median tensile strength from 35 MPa to 34 MPa. Then the probability of *obtaining* a statistically significant difference ($p < 0.05$) from the current median of 35 MPa for a sample of size 100 from the new furnace is only 0.21. However, this probability rises to 0.96 for a sample of size 1,000.

As a second illustration, assume that the new furnace reduces median tensile strength from 35 MPa to 31 MPa. Now the probability of *failing to obtain* a statistically significant difference ($p > 0.05$) from the current median is 0.01 for a sample of size 75 from the new furnace. However, for a sample of size 10, this probability is 0.72.

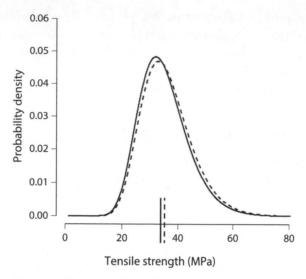

FIGURE 1. Assumed lognormal tensile strength distributions for specimens from the current (dashed curve) and the new (solid curve) furnaces. The median of each distribution is shown by a vertical line.

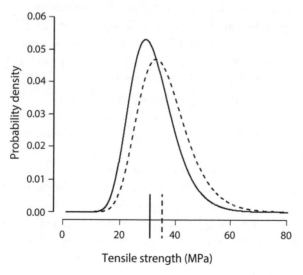

FIGURE 2. Alternative assumed lognormal tensile strength distributions for specimens from the current (dashed curve) and the new (solid curve) furnaces. The median of each distribution is shown by a vertical line.

These illustrations show how an effect of no practical importance can be deemed "statistically significant" for a sufficiently large sample, and how an effect that is of practical importance can (and is likely to) result in a "not statistically significant" call for an insufficient sample size.

Constructing a Statistical Interval

A statistical interval generally provides a more meaningful analysis than a significance test. In our first illustration, a simulated sample of 1,000 observations from the new furnace leads to a 95% confidence interval on median tensile strength from 33.6 MPa to 34.7 MPa. In our *QP* article, we said that:

> Roughly speaking, this means that one can be 95% sure that the median tensile strength for the new furnace is between 33.6 MPa and 34.7 MPa. More precisely, one can assert that if there were many such intervals calculated from different sets of random samples, about 95% of such intervals would, in fact, include the new median.

We also explained that, because the 95% confidence interval does not include 35 MPa, the median tensile strength for the new furnace differs significantly (at the 5% significance level, or $p < 0.05$) from its previously established value. We assert, however, that the confidence interval is more informative than the significance test. It shows that the deviation of median tensile strength from 35 MPa could be as small as 0.3 MPa (35–34.7) and is unlikely to exceed 1.4 MPa (35–33.6). But even the latter deviation might not be of practical importance—despite its statistical significance.

In our second illustration, a simulated sample of 10 observations leads to a 95% confidence interval on median tensile strength for the new furnace, from 26.6 MPa to 37.5 MPa. Because this interval includes 35 MPa, the median tensile strength for the new furnace does not differ significantly (at the 5% significance level) from its previous value. The confidence interval, however, suggests that the new furnace's median tensile strength could appreciably exceed 35 MPa (i.e., be as high as 37.5 MPa) or fall appreciably below (i.e., be as low as 26.6 MPa). Either of these conclusions—though in opposite directions—would be of practical importance. Thus, the real conclusion from our analysis is that a study with only 10 observations is insufficient to provide conclusive results, and a larger sample is required.

The Big Picture

We need to make clear to practitioners that statistical inference, irrespective of the specific approach, deals with drawing conclusions about a population or process based upon a random sample. The resulting statistical analyses reflect only the uncertainty due to sampling and exclude added uncertainty (or bias) due to the sample not being random.

Consider, for example, a TV news network that invites its audience to phone in its views on a political issue. The self-selected respondents are far from a random sample from the general public—typically the population of interest—and are likely to have views that are unrepresentative of that population. Constructing confidence intervals in such situations can lead to false trust in the findings. Thus, a key contribution of a statistical perspective is in planning the study *in advance* to ensure that the selected sample is as close as possible—given practical constraints—to a random sample from the population of interest. Our earlier illustrations, moreover, show the impact of sample size on the length of the constructed statistical intervals, and thus, the precision of the findings. Such assessments can, and should, be made prior to conducting the study to ensure that the resources required to obtain results with the desired precision are available.

While demonstrating the important role of statistical intervals in many analyses, we also need to discourage practitioners from a singular focus on such intervals as the exclusive pinnacle of their analyses, as has tended to be the case for *p*-values. There are numerous further important considerations in the design, execution, and analysis of statistical studies that we need to convey. Additionally, we need to convince practitioners of the importance of appropriate graphical displays of their data—often as a first step in the analyses.

Also, while confidence intervals are the most frequently used and discussed type of statistical interval, some practical applications call for other kinds of intervals—in particular, tolerance intervals to contain a specified proportion of a population, or prediction intervals to contain one or more future observations or other quantity of interest.[4]

Indeed, the specific use of a confidence (or other statistical) interval—or some other approach—to replace current inference procedures is typically application-dependent. But an advantage of using statistical intervals, when applicable, is that they use concepts with which practitioners tend to be familiar.

In summary, we need to convince practitioners that there is an important place for statistical inference beyond significance testing and $p < 0.05$. We hope the approach outlined here will help get this message across—and quickly.

References

1. Wasserstein, R. L., Schirm, A. L., and Lazar, N. A. (eds.) (2019) Statistical Inference in the 21st Century: A World beyond $p < 0.05$. *American Statistician*, **73**(sup1), 1–401.
2. Hahn, G. J. (1974) Don't Let Statistical Significance Fool You! *Chemtech*, **4**(1), 16–18.
3. Doganaksoy, N., Hahn, G. J., and Meeker, W. Q. (2017) Fallacies of Statistical Inference: The Case for Focusing Your Data Analyses beyond Significance Tests. *Quality Progress*, **51**(11), 56–62.
4. Meeker, W. Q., Hahn, G. J., and Escobar, L. A. (2017) *Statistical Intervals: A Guide for Practitioners and Researchers*. Hoboken, NJ: John Wiley & Sons.

Contributors

Colin Adams (https://orcid.org/0000-0002-9244-8368) is the Thomas T. Read Professor of Mathematics at Williams College, having received his Ph.D. from the University of Wisconsin-Madison in 1983. He is the author or co-author of numerous research papers and nine books, including *The Knot Book*, *How to Ace Calculus: The Streetwise Guide*, the comic book *Why Knot?*, the novel *Zombies & Calculus,* and various other texts. He is a recipient of the Haimo National Distinguished Teaching Award, a Pólya Lecturer for the Mathematical Association of America, a Sigma Xi Distinguished Lecturer, and a recipient of the Robert Foster Cherry Teaching Award. He appears in the videos "The Great pi/e Debate" and "Derivative vs. Integral: The Final Smackdown." He is also the humor columnist for *The Mathematical Intelligencer* and puts on humorous math theater at the Joint Mathematics Meetings every year.

John Baez did his undergraduate work at Princeton and did his Ph.D. in mathematics at MIT, where he worked with Irving Segal on quantum field theory. Since 1989, he has been teaching at the University of California, Riverside. He has worked on quantum gravity, *n*-categories, and more recently applied category theory. He enjoys blogging about mathematics and physics.

Bruce M. Boghosian (http://orcid.org/0000-0003-3040-3399) is Professor of Mathematics at Tufts University, where he also holds secondary positions in the departments of Physics and Computer Science, and he was selected as a Fellow of the Tisch College of Civic Life (2018–2019). His research is on applied dynamical systems and probability theory, with emphasis on mathematical kinetic theory. He is a Fellow of the American Physical Society (2000), was named a Distinguished Scholar of Tufts University (2010), and sits on the editorial boards of, inter alia, the *Journal of Computational Science* (since 2009), and *Physica A* (since 2001). He has held visiting research positions at, inter alia, the Université de Paris-Sud Orsay (2016 and 2019), and the École Normale Supérieure de Paris (2008). Boghosian also has strong interests in international higher education, having served as the President of the American University of Armenia, an affiliate of the University of California, for four years (2010–2014). In connection with that service, he was elected to the Armenian National Academy of Sciences (2008), and received the Order of the Republic of Armenia from that country's prime minister (2014).

Mark Colyvan is a professor of philosophy at the University of Sydney and a visiting professor at the Munich Center for Mathematical Philosophy, Ludwig Maximilian University, Munich. He is a past president of the Australasian Association of Philosophy and also a past president of the Society for Risk Analysis (Australia and New Zealand). His publications include *The Indispensability of Mathematics* (Oxford University Press, 2001), *An Introduction to the Philosophy of Mathematics* (Cambridge University Press, 2012), and, with Lev Ginzburg, *Ecological Orbits: How Planets Move and Populations Grow* (Oxford University Press, 2004).

Peter J. Denning (http://orcid.org/0000-0001-5278-2305) is a Distinguished Professor of Computer Science and director of the Cebrowski Institute for Information Innovation at the Naval Postgraduate School in Monterey, California. His research interests include innovation principles, great principles of computing, operating systems, performance modeling, workflow management, high-performance computing, security, data science, artificial intelligence, and computing education. He was a pioneer in virtual memory performance, performance evaluation of computer systems, and the early Internet. He has published more than 425 papers and 11 books, including most recently *Computational Thinking* and *Great Principles of Computing*. Denning received a Ph.D. in electrical engineering from MIT. He is editor of the Association for Computing Machinery's *Ubiquity* and a past president of ACM.

Necip Doganaksoy, Ph.D., (https://orcid.org/0000-0002-8641-0814) is the Douglas T. Hickey Chair in business and associate professor at Siena College School of Business in Loudonville, New York, following a 26-year career in industry, mostly at General Electric (GE). He has published extensively based on his research and applications in reliability, quality, and productivity improvement in business and industry. He is a Fellow of the American Statistical Association and the American Society for Quality. He is the co-author of *The Role of Statistics in Business and Industry* and *A Career in Statistics: Beyond the Numbers*, both published by Wiley.

Gerald J. Hahn, Ph.D., retired from what is now the GE Global Research Center after 46 years as corporate statistician, including 28 years as founder and manager of the 16-person Applied Statistics Program. He has been on the firing line in developing, applying, and explaining statistical methods to address key company problems, many dealing with assessing the impact on performance and reliability of a product change. He is author or co-author of five Wiley books and numerous statistical and engineering journal articles. Recipient of many awards, he is a Fellow of the American Society for Quality and the American Statistical Association.

Jim Henle is the Myra M. Sampson Professor Emeritus at Smith College. He is the author of more than a hundred papers, essays, columns, and books. With his son, Fred Henle, he won the David P. Robbins Prize of the MAA for their research paper, "Squaring the Plane." He has written or co-written texts in Calculus (*Calculus: The Language of Change,* with D. Cohen), Logic (*Sweet Reason: A Field Guide to Modern Logic,* with J. Garfield and T. Tymoczko), Nonstandard Analysis (*Infinitesimal Calculus,* with E. M. Kleinberg), and Set Theory (*An Outline of Set Theory*). In books and papers, Henle has argued variously that mathematics is poetry, music, language, gastronomy, fiction, logic, and art. He currently writes for *The Mathematical Intelligencer.* His most recent book is *The Proof and the Pudding: What Mathematicians, Cooks, and You Have in Common.* His latest project is a book of baseball puzzles.

Patrick Honner teaches mathematics at Stuyvesant High School in New York City. As a public school teacher, Patrick has taught everything from introductory algebra to multivariable calculus and has developed innovative courses in mathematical writing, research, and computing. He is a four-time recipient of *Math for America*'s Master Teacher Fellowship, a New York State Master Teacher, a National STEM Teacher Ambassador, and a recipient of the Presidential Award for Excellence in Mathematics and Science Teaching. He writes a column for *Quanta Magazine* that connects cutting edge research to curricular mathematics, and his work appears in publications like *WIRED* and *The New York Times.* You can learn more at PatrickHonner.com and follow him on Twitter at @MrHonner.

Chris King (http://dhushara.com) is an emeritus lecturer in mathematics from the University of Auckland, New Zealand, pursuing research in chaotic and fractal dynamics, including open source applications spanning chaotic dynamics, cellular automata, and Kleinian limit sets. He maintains a research web resource on biocosmology—how cumulative fractal quantum interactions arising from cosmic symmetry-breaking have led to the emergence and evolution of life, culminating in neurodynamic research into the foundations of subjective consciousness. He has complemented these with field trips to the sources of the world's entheogenic species, and to document the biodiversity crisis facing the Amazon. He maintains a cumulative web resource on biocrisis and is joint author of *Sexual Paradox: Complementarity, Reproductive Conflict and Human Emergence*, available from the website http://dhushara.com.

Erica Klarreich has been writing about mathematics and science for more than 20 years. She has a doctorate in mathematics from Stony Brook University and is a graduate of the Science Communication Program at the University of California, Santa Cruz. A freelance journalist based in Berkeley, California, her work has appeared in many publications, including *Quanta Magazine, Nature, The Atlantic, New Scientist, Nautilus*, and *Science News*, for which she was the mathematics correspondent for several years. She has been the Journalist in Residence at both the Mathematical Sciences Research Institute and the Simons Institute for the Theory of Computing in Berkeley. Her work also appeared in the 2010, 2011, and 2016 volumes of *The Best Writing on Mathematics*. For more of her writing, see www.ericaklarreich.com or follow her on Twitter: @EricaKlarreich

Ted G. Lewis (http://orcid.org/0000-0003-3467-0312) is an author, speaker, and consultant with expertise in applied complexity theory, homeland security, infrastructure systems, and early-stage startup strategies. He has served in government, industry, and academia over a long career, including as Executive Director and Professor of Computer Science, Center for Homeland Defense and Security, at the Naval Postgraduate School in Monterey, California; Senior Vice President of Eastman Kodak; President and CEO of DaimlerChrysler Research and Technology, North America; and Professor of Computer Science at Oregon State University, Corvallis. In addition, he has served as the Editor-in-Chief of a number of periodicals: *IEEE Computer Magazine, IEEE Software Magazine*, as a member of the IEEE Computer Society Board of Governors, and is currently Advisory Board Member of ACM's *Ubiquity* web-based magazine and *Cosmos+Taxis* journal (The Sociology of Hayek). He has published more than 30 books, most recently including *The Signal: The History of Signal Processing and How We Communicate, Book of Extremes: The Complexity of Everyday Things, Bak's Sand Pile: Strategies for a Catastrophic World, Network Science: Theory and Practice*, and *Critical Infrastructure Protection in Homeland Security: Defending a Networked Nation*. Lewis has authored or co-authored numerous scholarly articles in cross-disciplinary journals such as *Cognitive Systems Research, Homeland Security Affairs Journal, Journal of Risk Finance, Journal of Information Warfare, Communications of the ACM*,

American Scientist, and *IEEE Parallel & Distributed Technology*. Lewis resides with his wife and dog in Monterey, California.

Dave Linkletter is a Ph.D. candidate in pure mathematics at the University of Nevada, Las Vegas. He is researching set theory and mathematical foundations, with a dissertation in large cardinals to be presented in 2021. He earned his BA in mathematics from The College of New Jersey in 2013 and his MS in pure mathematics from DePaul University in 2015. His first published writing on mathematics was in 2019, with *Plus* Magazine, before he joined *Popular Mechanics* magazine, where he has written more than 10 articles thus far. After finishing his Ph.D., he aims to join a university where he can teach, further research on the frontier of large cardinals, and continue writing mathematically for all audiences.

William Q. Meeker, Ph.D. (https://orcid.org/0000-0002-5366-0294), is Distinguished Professor of Statistics at Iowa State University. He is a Fellow of the American Statistical Association, the American Society for Quality, and the American Association for the Advancement of Science. He is a past Editor of *Technometrics* and has won numerous awards for his research and publications. He has done research and consulted extensively on problems in reliability data analysis, warranty analysis, reliability test planning, accelerated testing, nondestructive evaluation, and statistical computing. He is the co-author of *Statistical Methods for Reliability Data* and the Second Edition of *Statistical Intervals*, both published by Wiley.

Richard Montgomery (https://orcid.org/0000-0001-5293-6837) works in celestial mechanics and sub-Riemannian geometry. He was led into both by the problem of the falling cat: how does a cat, falling from upside down with zero angular momentum, change her shape so as to land on her feet, and what is an optimal shape-changing maneuver? Before being a mathematician, Richard was known for first descents in California: going down difficult stretches of whitewater rivers for the first time, often in remote canyons, in a hard-shell kayak; he did most of those descents with Lars Holbek and Chuck Stanley, from 1978 to 1984.

Boris Odehnal (https://orcid.org/0000-0002-7265-5132) graduated from TU Wien (Vienna University of Technology, Austria) and received a Ph.D. in geometry and an M.Sc. in geometry and mathematics. He was a teaching assistant at the University of Natural Resources and Life Sciences Vienna and an assistant professor at the TU Wien, where he received his habilitation in geometry. He served as a full professor of geometry at the TU Dresden (Saxony, Germany) before he returned to Vienna, where he is now assistant professor at the University of Applied Arts Vienna. He is a coauthor of *The Universe of Conics* (2016) and *The Universe of Quadrics* (2020) (both Springer; with G. Glaeser and H. Stachel) and dozens of scientific articles. He has taught descriptive geometry, differential geometry, line geometry, kinematics, projective and elementary geometry, and mathematics among other subjects. His research interests lie in geometry and everything related to it.

Ben Orlin is a teacher and the creator of the blog "Math with Bad Drawings." His books *Math with Bad Drawings: Illuminating the Ideas That Shape Our Reality* (2018) and *Change is the Only Constant: The Wisdom of Calculus in a Madcap World* (2019) present mathematical ideas for a general audience. His writing has appeared in *Popular Science, The Atlantic, Slate, Vox, The Los Angeles Times, Ars Technica*, and elsewhere. He has taught math in the United States and the United Kingdom to every age from 11 to 18.

James Propp is a full professor in the Department of Mathematical Sciences at the University of Massachusetts Lowell. Most of his research is in combinatorics, probability, and dynamical systems theory, with forays into the study of cellular automata and games. His monthly essays are posted at his *Mathematical Enchantments* blog at http://mathenchant.wordpress.com; "Who Mourns the Tenth Heegner Number?" was one of those essays, as was "Prof. Engel's Marvelously Improbable Machines," which appeared in *The Best Writing on Mathematics 2019*. Propp is a member of the advisory council of the National Museum of Mathematics and the chair of the advisory council of the Gathering 4 Gardner Foundation. You can follow him on Twitter at @jimpropp.

Steven Strogatz (http://orcid.org/0000-0003-2923-3118) is the Jacob Gould Schurman Professor of Applied Mathematics at Cornell University. He works on nonlinear dynamics and complex systems applied to physics, biology, and the social sciences. According to Google Scholar, Strogatz's 1998 paper "Collective dynamics of small-world networks," co-authored with his former student Duncan Watts, ranks among the top 100 most-cited scientific papers of all time. His latest book, *Infinite Powers*, was a *New York Times* bestseller and was short-listed for the 2019 Royal Society Science Book Prize. Follow him on Twitter at @stevenstrogatz.

Donald Teets received his Doctor of Arts degree from Idaho State University in 1988 and has been a professor of mathematics at the South Dakota School of Mines and Technology since then. His research interests lie in the intersection of mathematics, history, and astronomy, particularly in the works of Gauss, Laplace, Lagrange, Bessel, and their contemporaries. His paper, "The Discovery of Ceres: How Gauss Became Famous" (*Mathematics Magazine* vol. 72, 1999, coauthor K. Whitehead) received the Carl B. Allendoerfer Award from the Mathematical Association of America in 2000. In addition to mathematics, his interests include bicycling, backpacking, rock climbing, and other outdoor pursuits.

Paul Thagard (paulthagard.com) is Distinguished Professor Emeritus of Philosophy at the University of Waterloo and Fellow of the Royal Society of Canada, the Cognitive Science Society, and the Association for Psychological Science. His books include *The Cognitive Science of Science: Explanation, Discovery, and Conceptual Change* (MIT Press 2012); *The Brain and the Meaning of Life* (Princeton University Press 2010); *Hot Thought: Mechanisms and Applications of Emotional Cognition* (MIT Press 2006); and *Mind: Introduction to Cognitive Science* (MIT Press 1996; second edition, 2005). His Treatise on Mind and Society series was published by Oxford University Press in 2019: *Brain-Mind*, *Mind-Society*, and *Natural Philosophy*.

Jørgen Veisdal (https://orcid.org/0000-0001-7325-0607) holds B.A. and M.Sc. degrees in engineering, a B.Sc. in mathematics, and is currently a Ph.D. research fellow at the Norwegian University of Science and Technology. His current research regards entry strategies for platform businesses in multisided markets and has been published in *Electronic Markets*. Day-to-day, he is also editor of the digital magazine *Cantor's Paradise,* which publishes essays on mathematics-related topics. His weekly essays, including "The Unparalleled Genius of John von Neumann," "The Riemann Hypothesis, Explained," and "The Beautiful Life of John Forbes Nash, Jr." have since their publication received more than 1.5 million views on the digital publishing platform Medium.

Stan Wagon (http://stanwagon.com/) (https://orcid.org/0000-0002-4524-0767) obtained his Ph.D. in set theory at Dartmouth College in 1975 and is now retired from teaching, having taught at Smith and Macalester Colleges. He has won several writing awards including the Trevor Evans award for the paper in this collection. His books include *Which Way Did the Bicycle Go?* and *The Banach–Tarski Paradox*. His construction of a full-sized square-wheeled bicycle attracted a lot of attention and earned him an entry in *Ripley's Believe It or Not*. He has been a runner and long-distance skier for many years and was a founding editor of *Ultrarunning* magazine.

Notable Writings

As a space-saving rule, the two lists that follow are complementary; that is, I did not include in the first list the titles of pieces published in the special journal issues mentioned on the second list. Because of time and other resource constraints related to the spring 2020 health crisis, I did not cover as much literature as I would have liked. Even in normal times, some periodicals are available to me only in paper copy in the libraries at Cornell University and Syracuse University; with both libraries closed while I worked on this section, my bibliographic research was negatively impacted, limited to electronic resources. For the same reasons, I failed to put together other lists I usually offer to the readers, on significant book reviews, notable teaching tips, and memorial notes.

In some electronic-only journals, the pagination of articles no longer starts with the page number that follows the last page of previous articles—at least in the versions I was able to download. In such cases, I gave the pagination starting with page one and ending with the page number equal to the number of pages of the piece. Thus, in the first list it is possible to find different entries that apparently overlap in reference location.

Acerbi, F. "There Is No *Consequentia Mirabilis* in Greek Mathematics." *Archive for History of Exact Sciences* 73(2019): 217–42.

Almira, José María, José Ángel Cid, and Julio Ostalé. "When Did Hermann Weyl Pass Away?" *British Journal for the History of Mathematics* 34.1(2019): 60–63.

Anatriello, Giuseppina, and Giovanni Vincenzi. "On the Definition of Periodic Decimal Representations: An Alternative Point of View." *The Mathematics Enthusiast* 16.1(2019): 3–14.

Anglade, Marie, and Jean-Yves Briend. "Le Diamètre et la Traversale: Dans l'Atelier de Girard Desargues." *Archive for History of Exact Sciences* 73(2019): 385–426.

Arndt, Michael. "The Role of Structural Reasoning in the Genesis of Graph Theory." *History and Philosophy of Logic* 40.3(2019): 266–97.

Arney, Chris, and Ammanda Beecher. "Complexity Modeling: Solving Modern Problems and Confronting Real Issues." *UMAP Journal* 40.1(2019): 71–84.

Barahmand, Ali. "On Mathematical Conjectures and Counterexamples." *Journal of Humanistic Mathematics* 9.1(2019).

Bardi, Alberto. "The Reception and Rejection of 'Foreign' Astronomical Knowledge in Byzantium." *Finding, Inheriting, or Borrowing? The Construction and Transfer of Knowledge in Antiquity and the Middle Ages*. Bielefeld, Germany: Transcript Verlag, 2019, 167–84.

Bartlett, Christopher. "Nautilus Spirals and the MetaGolden Ratio Chi." *Nexus Network Journal* 21(2019): 541–56.

Belyaev, Alexander, and Pierre-Alain Fayolle. "Counting Parallel Segments: New Variants of Pick's Area Theorem." *The Mathematical Intelligencer* 41.4(2019): 1–7.

Biggs, Norman. "John Reynolds of the Mint: A Mathematician in the Service of King and Commonwealth." *Historia Mathematica* 48(2019): 1–28.

Biggs, Norman. "Thomas Harriot on the Coinage of England." *Archive for History of Exact Sciences* 73(2019): 361–83.

Blåsjö, Viktor, and Jan P. Hogendijk. "On Translating Mathematics." *Isis* 109.4(2018): 774–81.

Boddy, Rachel. "Frege's Unification." *History and Philosophy of Logic* 40.2(2019): 135–51.

Boi, Luciano. "Some Mathematical, Epistemological, and Historical Reflections on the Relationship between Geometry and Reality, Space–Time Theory and the Geometrization of Theoretical Physics, from Riemann to Weyl and Beyond." *Foundations of Science* 24(2019): 1–38.

Braithwaite, David W., et al. "Individual Differences in Fraction Arithmetic Learning." *Cognitive Psychology* 112(2019): 81–98.

Braun, Benjamin, and Eric Kahn. "Teaching History of Mathematics: A Dialogue." *Journal of Humanistic Mathematics* 9.1(2019).

Brown, Aaron. "Kelly Attention." *Wilmott Magazine* 99(2019): 8–11.

Brunheira, Lina, and João Pedro da Ponte. "From the Classification of Quadrilaterals to the Classification of Prisms." *The Journal of Mathematical Behavior* 53(2019): 65–80.

Bubelis, William S. "Imperial Numeracy? Athenian Calculation of Some Imperial Taxes." *Hegemonic Finances: Funding Athenian Domination in the 5th and 4th Centuries BC*, edited by Thomas J. Figueira and Sean R. Jensen. Swansea, U.K.: The Classical Press of Wales, 2019, 25–54.

Bueno, Otávio. "Structural Realism, Mathematics, and Ontology." *Studies in History and Philosophy of Science, A* 74(2016): 4–9.

Bussotti, Paolo. "Michel Chasles' Foundational Programme for Geometry until the Publication of His Aperçu Historique." *Archive for History of Exact Sciences* 73(2019): 261–308.

Caglayan, Günhan. "Theory of Polygonal Numbers with Cuisenaire Rods Manipulatives: Understanding Theon of Smyrna's Arithmetic in a History of Mathematics Classroom." *British Journal for the History of Mathematics* 34.1(2019): 12–22.

Caliò, Franca, and Elena Marchetti. "Two Architectures: Two Compared Geometries." *Nexus Network Journal* 21(2019): 527–45.

Campbell, Julianna, and Christine von Renesse. "Learning to Love Math through the Exploration of Maypole Patterns." *Journal of Mathematics and the Arts* 13.1–2(2019): 131–51.

Campbell, Paul J. "The Coming Great Pall of China: Demography of a One-Child Family." *UMAP Journal* 40.1(2019): 47–70.

Campbell, Stephen Maxwell. "Exponentials and Logarithms: A Proposal for a Classroom Art Project." *Journal of Mathematics and the Arts* 13.1–2(2019): 91–99.

Camúñez-Ruiz, José Antonio, and María Dolores Pérez-Hidalgo. "Juan Caramuel (1606–1682) and the Spanish Version of the Passedix Game." *British Journal for the History of Mathematics* 34.3(2019): 143–54.

Carman, Christián C., and Gonzalo L. Recio. "Ptolemaic Planetary Models and Kepler's Laws." *Archive for History of Exact Sciences* 73(2019): 39–124.

Carter, Jessica. "Exploring the Fruitfulness of Diagrams in Mathematics." *Synthese* 196(2019): 4011–32.

Carter, Jessica. "Philosophy of Mathematical Practice: Motivations, Themes and Prospects." *Philosophia Mathematica* 27(2019): 1–32.

Cellucci, Carlo. "Diagrams in Mathematics." *Foundations of Science* 24(2019): 583–604.

Chabas, José, and Bernard R. Goldstein. "The Medieval Moon in a Matrix; Double Argument Tables for Lunar Motion." *Archive for History of Exact Sciences* 73(2019): 335–59.

Chassé, Daniel Spieh. "In Search for a Global Centre of Calculation: The Washington Statistical Conferences of 1947." *The Force of Comparison: A New Perspective on Modern European History and the Contemporary World*. New York: Berghahn, 2019, 266–87.

Cheng, Diana, Tetyana Berezovski, and Rachael Talbert. "Dancing on Ice: Mathematics of Blade Tracings." *Journal of Mathematics and the Arts* 13.1–2(2019): 112–30.

Chow, Timothy Y. "The Consistency of Arithmetic." *The Mathematical Intelligencer* 41.1(2019): 23–30.

Chu, Longfei, and Haohao Zhu. "Re-examining the Impact of European Astronomy in Seventeenth-Century China: A Study of Xue Fengzuo's System of Thought and His Integration of Chinese and Western Knowledge." *Annals of Science* 76.3–4(2019): 303–23.

Claveau, François, and Olivier Grenier. "The Variety-of-Evidence Thesis: A Bayesian Exploration of Its Surprising Failures." *Synthese* 196(2019): 3001–28.

Connera, AnnaMarie, and Carlos Nicolas Gomez. "Belief Structure as Explanation for Resistance to Change: The Case of Robin." *The Journal of Mathematical Behavior* 56(2019): 196–201.

Copeland, B. Jack, and Oron Shagrir. "The Church-Turing Thesis: Logical Limit or Breachable Barrier?" *Communications of the ACM* 62.1(2019): 66–74.

Crippa, Davide, and Pietro Milici. "A Relationship between the Tractrix and Logarithmic Curves with Mechanical Applications." *The Mathematical Intelligencer* 41.4(2019): 29–34.

Cui, Jiaxin, et al. "Visual Form Perception Is Fundamental for Both Reading Comprehension and Arithmetic Computation." *Cognition* 189(2019): 141–54.

Dagpunar, John. "The Gamma Distribution." *Significance* 16.1(2019): 10–11.

Dalal, Siddhartha, Berkay Adlim, and Michael Lesk. "How to Measure Relative Bias in Media Coverage?" *Significance* 16.5(2019): 18–23.

Dana-Picard, Th., and S. Hershkovitz. "Geometrical Features of a Jewish Monument." *Journal of Mathematics and the Arts* 13.1–2(2019): 60–71.

Davies, James E. "Towards a Theory of Singular Thought about Abstract Mathematical Objects." *Synthese* 196(2019): 4113–36.

De Jong, Teije. "A Study of Babylonian Planetary Theory." *Archive for History of Exact Sciences* 73(2019): 1–37 and 309–33.

De Las Peñas, Ma. Louise Antonette, and Analyn Salvador-Amores. "Enigmatic Geometric Tattoos of the 'Butbut' of Kalinga, Philippines." *The Mathematical Intelligencer* 41.1(2019): 31–38.

Dean, Walter. "Computational Complexity Theory and the Philosophy of Mathematics." *Philosophia Mathematica* 27(2019): 381–439.

Della Puppa, Giovanni, Roger A. Sauer, and Martin Trautz. "A Unified Representation of Folded Surfaces via Fourier Series." *Nexus Network Journal* 21(2019): 491–526.

Deshpande, Anjali, and Shannon Guglielm. "Four Moves to Motivate Students in Problem Solving." *The Mathematics Teacher* 112.7(2019): 510–15.

Devadoss, Satyan L., and Diane Hoffoss. "Unfolding Humanity: Mathematics at Burning Man." *Notices of the American Mathematical Society* 65.4(2019): 572–75.

Di Lazzaro, Paolo, Daniele Murra, and Pietro Vitelli. "The Interdisciplinary Nature of Anamorphic Images in a Journey through Art, History and Geometry." *Journal of Mathematics and the Arts* 13.4(2019): 353–68.

Díaz, Juan Pablo, Gabriela Hinojosa, Martha Mendoza, and Alberto Verjovsky. "Dynamically Defined Wild Knots and Othoniel's *My Way.*" *Journal of Mathematics and the Arts* 13(3): 230–42.

Dimmel, Justin, and Amanda Milewski. "Scale, Perspective, and Natural Mathematical Questions." *For the Learning of Mathematics* 39.3(2019): 34–40.

Domokos, Gábor. "The Gömböc Pill." *The Mathematical Intelligencer* 41.2(2019): 8–11.

Downey, Rod, and Denis R. Hirschfeldt. "Algorithmic Randomness." *Communications of the ACM* 62.5(2019): 70–80.

Dunér, David. "The Axiomatic-Deductive Ideal in Early Modern Thinking: A Cognitive History of Human Rationality." *Cognitive History: Mind, Space, and Time*, edited by David Dunér and Christer Ahlberger. Berlin: De Gruyer, 2019, 99–126.

Eggers, Michael. "Outlines of a Historical Epistemology of Comparison: From Descartes to the Early Nineteenth Century." *The Force of Comparison: A New Perspective on Modern European History and the Contemporary World.* New York: Berghahn, 2019, 33–52.

Eisenmann, Petr, and Martin Kuřil. "Number Series and Computer." *Mathematics Enthusiast* 16(2019): 253–62.

Ellis, Amy, Zekiye Özgür, and Lindsay Reiten. "Teacher Moves for Supporting Student Reasoning." *Mathematics Education Research Journal* 31(2019): 107–32.

Engelstein, Geoffrey. "When Math Doesn't Have All the Answers." *GameTek.* New York: HaperCollins, 2019, 170–87.

Evans, James, and Christián C. Carman. "Babylonian Solar Theory on the Antikythera Mechanism." *Archive for History of Exact Sciences* 73(2019): 619–59.

Fehér, Krisztina, et al. "Pentagons in Medieval Sources and Architecture." *Nexus Network Journal* 21(2019): 681–703.

Findlen, Paula. "The Renaissance of Science." *The Oxford Illustrated History of the Renaissance,* edited by Gordon Campbell. Oxford, U.K.: Oxford University Press, 2019, 378–425.

Fiss, Andrew, and Laura Kasson Fiss. "The Gömböc Pill." *Configuration* 27.3(2019): 301–29.

Franklin, James. "Mathematics, Core of the Past and Hope of the Future." *Reclaiming Education: Renewing Schools and Universities in Contemporary Western Culture,* edited by Catherine A. Runcie and David Brooks. Sydney, Australia: Edwin E. Lowe Publishing, 2018, 149–62.

Friberg, Jöran. "Three Thousand Years of Sexagesimal Numbers in Mesopotamian Mathematical Texts." *Archive for History of Exact Sciences* 73(2019): 183–216.

Friedman, Michael. "Mathematical Formalization and Diagrammatic Reasoning: The Case Study of the Braid Group between 1925 and 1950." *British Journal for the History of Mathematics* 34.1(2019): 43–59.

Fuchs, Wladek. "The Geometric Language of Roman Theater Design." *Nexus Network Journal* 21(2019): 547–90.

Fuentes, Paula. "The Islamic Crossed-Arch Vaults in the Mosque of Córdoba." *Nexus Network Journal* 21(2019): 441–63.

Gal, Ofer, and Cindy Hodoba Eric. "Between Kepler and Newton: Hooke's 'Principles of *Congruity* and *Incongruity*' and the Naturalization of Mathematics." *Annals of Science* 76.3–4(2019): 241–66.

Gao, Shan. "The Measurement Problem Revisited." *Synthese* 196(2019): 299–311.

Gauthier, Sébastien, and François Lê. "On the Youthful Writings of Louis J. Mordell on the Diophantine Equation $y^2 - k = x^3$." *Archive for History of Exact Sciences* 73.5(2019): 427–68.

German, Andy. "Mathematical Self-Ignorance and Sophistry: Theodorus and Protagoras." *Knowledge and Ignorance of Self in Platonic Philosophy,* edited by James M. Ambury and Andy German. Cambridge, U.K.: Cambridge University Press, 2019, 151–68.

Gibbs, Cameron. "Basing for the Bayesian." *Synthese* 196(2019): 3815–40.

Gilboa, Nava, Ivy Kidron, and Tommy Dreyfus. "Constructing a Mathematical Definition: The Case of the Tangent." *International Journal of Mathematical Education in Science and Technology* 50.3(2019): 421–46.

Glickman, Moshe, and Marius Usher. "Integration to Boundary Indecisions between Numerical Sequences." *Cognition* 193(2019): 1–8.

Grinbaum, Alexei. "The Effectiveness of Mathematics in Physics of the Unknown." *Synthese* 196(2019): 973–89.

Haas, Richard. "Maligned for Mathematics: Sir Thomas Urquhart and His *Trissotetras.*" *Annals of Science* 76.2(2019): 113–56.

Haffner, Emmylou. "From Modules to Lattices: Insight into the Genesis of Dedekind's *Dualgruppe.*" *British Journal for the History of Mathematics* 34.1(2019): 23–42.

Halimi, Brice. "Settings and Misunderstandings in Mathematics." *Synthese* 196(2019): 4623–56.

Hall, Andreia, Paulo Almeida, and Ricardo Teixeira. "Exploring Symmetry in Rosettes of Truchet Tiles." *Journal of Mathematics and the Arts* 13.4(2019): 308–35.

Hamdan, May. "Filling an Area Discretely and Continuously." *International Journal of Mathematical Education in Science and Technology* 50.8(2019): 1210–22.

Hanke, Miroslav. "Jesuit Probabilistic Logic between Scholastic and Academic Philosophy." *History and Philosophy of Logic* 40.4(2019): 355–73.

Harriss, E. O., C. Smith, and A. Carpenter. "Geometry in the Walnut Grove: An Applied Mathematical Approach to Art." *Journal of Mathematics and the Arts* 13.1–2(2019): 152–72.

Hart, K. P. "Machine Learning and the Continuum Hypothesis." *Nieuw Archief voor Wiskunde* 5/20.3(2019): 214–17.

Hartnett, Kevin. "With Category Theory, Mathematics Escapes from Equality." *Quanta Magazine* October 10, 2019. https://www.quantamagazine.org/with-category-theory-mathematics-escapes-from-equality-20191010/.

Haug, Espen Gaarder. "High-Speed Trading: The Trans-Atlantic Submarine Bridge." *Wilmott Magazine* 102(2019): 12–17.

Haug, Espen Gaarder. "Philosophy of Randomness: Time in Relation to Uncertainty." *Wilmott Magazine* 99(2019): 12–13.

Hawes, Zachary, et al. "Relations between Numerical, Spatial, and Executive Function Skills and Mathematics Achievement." *Cognitive Psychology* 109(2019): 68–90.

Hawthorne, Casey, and Bridget K. Druken. "Looking for and Using Structural Reasoning." *The Mathematics Teacher* 112.4(2019): 294–301.

Hecht, Eugene. "Kepler and the Origins of the Theory of Gravity." *American Journal of Physics* 87(2019): 176–85.

Hedman, Bruce A. "Cantor and the Infinity of God." *The Infinity of God: New Perspectives in Theology and Philosophy,* edited by Benedikt Paul Göcke and Christian Tapp. Notre Dame, IN: Notre Dame Press, 2019, 167–83.

Hohol, Mateusz, and Marcin Miłkowski. "Cognitive Artifacts for Geometric Reasoning." *Foundations of Science* 24(2019): 657–80.

Hou, Juncheng, and Saralees Nadaraja. "New Conditions for Independence of Events." *International Journal of Mathematical Education in Science and Technology* 50.2(2019): 322–24.

Houston-Edwards, Kelsey. "Numbers Game." *Scientific American* 321.3(2019): 35–40.

Hu, Qingfen, and Meng Zhang. "The Development of Symmetry Concept in Preschool Children." *Cognition* 189(2019): 131–40.

Hußmann, Stephan, Florian Schacht, and Maike Schindler. "Tracing Conceptual Development in Mathematics: Epistemology of Webs of Reasons." *Mathematics Education Research Journal* 31(2019): 133–49.

Ibarra, Lina Medina, Avenilde Romo-Vázquez, and Mario Sánchez Aguilar. "Using the Work of Jorge Luis Borges to Identify and Confront Students' Misconceptions about Infinity." *Journal of Mathematics and the Arts* 13.1(2019): 48–59.

Ippoliti, Emiliano. "Heuristics and Inferential Microstructures: The Path to Quaternions." *Foundations of Science* 24(2019): 411–25.

Iversen, Paul, and Alexander Jones. "The Back Plate Inscription and Eclipse Scheme of the Antikythera Mechanism Revisited." *Archive for History of Exact Sciences* 73(2019): 469–511.

Johns, Christopher. "The Impact of Leibniz's Geometric Method for the Law." *Leibniz's Legacy and Impact,* edited by Julia Weckend and Lloyd Strickland. New York: Routledge, 2019, 243–67.

Johnson, Ron, Kelvin Jones, and David Manley. "Why Geography Matters." *Significance* 16.1(2019): 32–37.

Johnson, Samuel G. B., and Stefan Steinerberger. "Intuitions about Mathematical Beauty: A Case Study in Aesthetic Experience of Ideas." *Cognition* 189(2019): 242–59.

Johnson, Samuel G. B., and Stefan Steinerberger. "The Universal Aesthetics of Mathematics." *The Mathematical Intelligencer* 41.1(2019): 67–70.

Jones, Chris, and Arthur Pewsey. "The sinh-arcsinh Normal Distribution." *Significance* 16.2(2019): 6–7.

Kalantari, Bahman, and Eric Hans Lee. "Newton–Ellipsoid Polynomiography." *Journal of Mathematics and the Arts* 13.4(2019): 336–52.

Kalinec-Craig, Crystal, Priya V. Prasad, and Carolyn Luna. "Geometric Transformations and Talavera Tiles: A Culturally Responsive Approach to Teacher Professional Development and Mathematics Teaching." *Journal of Mathematics and the Arts* 13.1–2(2019): 72–90.

Karaali, Gizem. "Doing Math in Jest: Reflections on Useless Math, the Unreasonable Effectiveness of Mathematics, and the Ethical Obligations of Mathematicians." *The Mathematical Intelligencer* 41.3(2019): 10–13.

Kinser-Traut, Jennifer Y. "Why Math?" *The Mathematics Teacher* 112.7(2019): 526–30.

Klarreich, Erica. "Out of a Magic Math Function, One Solution to Rule Them All." *Quanta Magazine* May 13, 2019. https://www.quantamagazine.org/universal-math-solutions-in -dimensions-8-and-24-20190513/.

Kleinman, Kim. "Why Edgar Anderson Visited Math Departments: Natural History, Statistics, and Applied Mathematics." *Historical Studies in the Natural Sciences* 49.1(2019): 41–69.

Kontorovich, Igor. "Non-Examples of Problem Answers in Mathematics with Particular Reference to Linear Algebra." *The Journal of Mathematical Behavior* 54(2019): 1–12.

Kozlowski, Joseph S., Scott A. Chamberlin, and Eric Mann. "Factors That Influence Mathematical Creativity." *Mathematics Enthusiast* 16(2019): 505–40.

Kwon, Oh Hoon, Ji-Won Son, and Ji Yeong I. "Revisiting Multiplication Area Models for Whole Numbers." *Mathematics Enthusiast* 16.1–3(2019): 359–68.

Lavers, Gregory. "Hitting a Moving Target: Gödel, Carnap, and Mathematics as Logical Syntax." *Philosophia Mathematica* 27(2019): 219–43.

Lazar, Nicole. "Big Data and Privacy." *Chance* 32.1(2019): 55–58.

Lazar, Nicole. "Crowdsourcing Your Way to Big Data." *Chance* 32.2(2019): 43–46.

Leemis, Lawrence M., and Raghu Pasupathy. "The Ties That Bind." *Significance* 16.4(2019): 8–9.

Lew, Kristen, and Dov Zazkis. "Undergraduate Mathematics Students' At-Home Exploration of a Prove-or-Disprove Task." *The Journal of Mathematical Behavior* 54(2019): 1–15.

Libertini, Jessica, and Troy Siemers. "Exposition." *UMAP Journal* 40.2–3(2019): 259–65.

Lipka, Jerry, et al. "Symmetry and Measuring: Ways to Teach the Foundations of Mathematics Inspired by Yupiaq Elders." *Journal of Humanistic Mathematics* 9.1(2019).

Lockwood, Elise, Anna F. DeJarnette, and Matthew Thomas. "Computing as a Mathematical Disciplinary Practice." *The Journal of Mathematical Behavior* 54(2019): 1–18.

Luecking, Stephen. "Visual Teaching of Geometry and the Origins of 20th Century Abstract Art." *Journal of Humanistic Mathematics* 9.2(2019).

Lützen, Jesper. "How Mathematical Impossibility Changed Welfare Economics: A History of Arrow's Impossibility Theorem." *Historia Mathematica* 46(2019): 56–87.

MacLean, Leonard, and Bill Ziemba. "The Efficiency of the NFL Betting Markets." *Wilmott Magazine* 99(2019): 30–34.

Mahajan, Sanjoy. "Don't Demean the Geometric Mean." *American Journal of Physics* 87.1(2019): 75–77.

Maidment, Alison, and Mark McCartney. "'A Man Who Has Infinite Capacity for Making Things Go': Sir Edmund Taylor Whittaker (1873–1956)." *British Journal for the History of Mathematics* 34.3(2019): 179–93.

Makovicky, Emil. "Pseudoheptagonal Mosaic from the Hodja Ahmad Tomb, Samarkand." *Nexus Network Journal* 21(2019): 657–67.

Malone, Stephanie, et al. "Learning Correspondences between Magnitudes, Symbols and Words: Evidence for a Triple Code Model of Arithmetic Development." *Cognition* 187 (2019): 1–9.

Maor, Eli. "The Chords of the Universe." *Aeon* May 30, 2018. https://aeon.co/essays/ringing -the-chords-of-the-universe-how-music-influenced-science.

Marmur, Ofer. "Key Memorable Events: A Lenson Affect, Learning, and Teaching in the Mathematics Classroom." *The Journal of Mathematical Behavior* 54(2019): 1–16.

Maronne, Sébastien. "Descartes's Mathematics." *The Oxford Handbook of Descartes and Cartesianism*, edited by Steven Nadler, Tad M. Schmaltz, and Delphine Antoine-Mahut. Oxford, U.K.: Oxford University Press, 2019, 138–56.

McCloskey, Andrea, and Samuel Jaye Tanner. "Ritual and Improvisation: Ways of Researching, Ways of Being in Mathematics Classrooms." *For the Learning of Mathematics* 39.2(2019): 37–41.

McCune, David, Lori McCune, and Dalton Nelson. "The Cutoff Paradox in the Kansas Presidential Caucuses." *UMAP Journal* 40.1(2019): 21–45.

Mitchell, Daniel Jon. "'The Etherealization of Common Sense?' Arithmetical and Algebraic Modes of Intelligibility in Late Victorian Mathematics of Measurement." *Archive for History of Exact Sciences* 73(2019): 125–80.

Moore, A. W. "Bird on Kant's Mathematical Antinomies," [originally published in 2011–12]. *Language, World, and Limits: Essays in the Philosophy of Language and Metaphysics*. Oxford, U.K.: Oxford University Press, 2019.

Moore, Kevin C., et al. "Conventions, Habits, and U.S. Teachers' Meanings for Graphs." *The Journal of Mathematical Behavior* 53(2019): 179–95.

Mount, Beau Madison. "Antireductionism and Ordinals." *Philosophia Mathematica* 27(2019): 105–24.

Moyon, Marc. "The *Liber Restauracionis*: A Newly Discovered Copy of a Medieval Algebra in Florence." *Historia Mathematica* 46(2019): 1–37.

Mozaffari, Mohammad S. "Ibn al-Fahhād and the Great Conjunction of 1166 AD." *Archive for History of Exact Sciences* 73(2019): 517–49.

Mundici, Daniele. "De Finetti Coherence and the Product Law for Independent Events." *Synthese* 196(2019): 265–71.

Najera, Jesus. "Unexpected Beauty in Primes." *Medium* October 12, 2019. https://medium.com/cantors-paradise/unexpected-beauty-in-primes-b347fe0511b2.

Naranan, S., T. V. Suresh, and Swarna Srinivasan. "Kolam Designs Based on Fibonacci Series." *Nieuw Archief voor Wiskunde* 5/20.3(2019): 195–204.

Nardelli, Enrico. "Do We Really Need Computational Thinking?" *Communications of the ACM* 62.2(2019): 32–35.

Nestler, Scott, and Andrew Hall. "The Variance Gamma Distribution." *Significance* 16.5 (2019): 10–11.

Neves, Juliano C. S. "Infinities as Natural Places." *Foundations of Science* 24(2019): 39–49.

Nickel, Gregor, "Aspects of Freedom in Mathematical Proof." *Zentralblatt für Didaktik der Mathematik* 51(2019): 845–56.

Niemi, Laura, et al. "Partisan Mathematical Processing of Political Polling Statistics: It's the Expectations That Count." *Cognition* 186(2019): 95–107.

Niss, Mogens. "The Very Multi-Faceted Nature of Mathematics Education Research." *For the Learning of Mathematics* 39.2(2019): 2–7.

Öhman, Lars-Daniel. "Are Induction and Well-Ordering Equivalent?" *The Mathematical Intelligencer* 41.3(2019): 33–40.

Oktaç, Asuman, María Trigueros, and Avenilde Romo. "APOS Theory: Connecting Research and Teaching." *For the Learning of Mathematics* 39.1(2019): 33–37.

Orlova, Nadezda, and Sergei Soloviev. "Logic and Logicians in Russia before 1917: Living in a Wider World." *Historia Mathematica* 46(2019): 38–55.

Overman, Kerenleigh A. "Materiality and the Prehistory of Number." *Squeezing Minds from Stones: Cognitive Archaeology and the Evolution of the Human Mind*, edited by Kerenleigh A. Overman and Frederick L. Coolidge. Oxford, U.K.: Oxford University Press, 2019, 432–56.

Palatnik, Alik, and Boris Koichu. "Flashes of Creativity." *For the Learning of Mathematics* 39.2(2019): 8–12.

Papadopoulos, Athanase. "Maps with Least Distortion between Surfaces." *Notices of the American Mathematical Society* 65.10(2019): 1628–39.

Papineau, David. "Knowledge Is Crude." *Aeon* June 3, 2019. https://aeon.co/essays/knowledge-is-a-stone-age-concept-were-better-off-without-it.

Pearl, Judea. "The Seven Tools of Causal Inference, with Reflections on Machine Learning." *Communications of the ACM* 62.3(2019): 54–60.

Peltonen, Kirsi. "Sensual Mathematics." *Journal of Mathematics and the Arts* 13.1–2(2019): 185–210.

Pinto, Alon. "Variability in the Formal and Informal Content Instructors Convey in Lectures." *The Journal of Mathematical Behavior* 54(2019): 1–17.

Pinto, Mario, et al. "Reconstructing the Origins of the Space-Number Association." *Cognition* 190(2019): 143–56.

Portaankorva-Koivisto, Päivi, and Mirka Havinga. "Integrative Phenomena in Visual Arts and Mathematics." *Journal of Mathematics and the Arts* 13.1–2(2019): 4–24.

Proulx, Jérôme. "Mental Mathematics under the Lens: Strategies, Oral Mathematics, Enactments of Meanings." *The Journal of Mathematical Behavior* 56(2019): 1–16.

Quinn, Candice M., et al. "Music as Math Waves: Exploring Trigonometry through Sound." *Journal of Mathematics and the Arts* 13.1–2(2019): 173–84.

Raper, Simon. "Fisher's Random Idea." *Significance* 16.1(2019): 20–22.

Raper, Simon. "Pearson and the Parameter." *Significance* 16.3(2019): 22–25.

Raynaud, Dominique, Samuel Gessner, and Bernardo Mota. "Andalò di Negro's *De compositione astrolabii*: A Critical Edition with English Translation and Notes." *Archive for History of Exact Sciences* 73(2019): 551–617.

Reichenberger, Andrea. "From Solvability to Formal Decidability: Revisiting Hilbert's 'Non-Ignorabimus.'" *Journal of Humanistic Mathematics* 9.1(2019).

Reinke, Luke T. "Toward an Analytical Framework for Contextual Problem-Based Mathematics Instruction." *Mathematical Thinking and Learning* 21.4(2019): 265–84.

Rekvenyi, Kamilla. "Paul Erdős's Mathematics as a Social Activity." *British Journal for the History of Mathematics* 34.2(2019): 134–42.

Reynolds, Mark A., and Stephen R. Wassell. "'Marriage of Incommensurables': A Geometric Conversation between an Artist and a Mathematician." *Journal of Mathematics and the Arts* 13.3(2019): 211–29.

Richardson, John T. E. "Who Introduced Western Mathematicians to Latin Squares?" *British Journal for the History of Mathematics* 34.2(2019): 95–103.

Richman, Andrew S., Leslie Dietiker, and Meghan Riling. "The Plot Thickens: The Aesthetic Dimensions of a Captivating Mathematics Lesson." *The Journal of Mathematical Behavior* 54(2019): 1–15.

Rockmore, Dan. "The Myth and Magic of Generating New Ideas." *The New Yorker* November 7, 2019. https://www.newyorker.com/culture/annals-of-inquiry/the-myth-and-magic-of-generating-new-ideas.

Rodríguez-Vellando, Pablo. ". . . And so Euler *Discovered* Differential Equations." *Foundations of Science* 24(2019): 343–74.

Ross, Helen Elizabeth, and Betty Irene Knott. "Dicuil (9th Century) on Triangular and Square Numbers." *British Journal for the History of Mathematics* 34.2(2019): 79–94.

Rossini, Paolo. "New Theories for New Instruments: Fabrizio Mordente's Proportional Compass and the Genesis of Giordano Bruno's Atomist Geometry." *Studies in History and Philosophy of Science, A* 76(2016): 60–68.

Rowe, David E. "On Emmy Noether's Role in the Relativity Revolution." *The Mathematical Intelligencer* 41.2(2019): 65–72.

Rowlett, Peter, et al. "The Potential of Recreational Mathematics to Support the Development of Mathematical Learning." *International Journal of Mathematical Education in Science and Technology* 50.7(2019): 972–86.

Ruagh, Michael, and Siegmund Probst. "The Leibniz Catenary and Approximation of *e*: An Analysis of His Unpublished Calculations." *Historia Mathematica* 49(2019): 1–19.

Russell, Craig. "Connecting Mathematics with World Heritage." *The Mathematics Teacher* 112.4(2019): 274–79.

Ryan, Ulrika. "Mathematical Preciseness and Epistemological Sanctions." *For the Learning of Mathematics* 39.2(2019): 25–29.

Sābetghadam, Zahrā. "Improving Spoke Wheel Roofs with the Geometry of Rasmi-Bandis." *Nexus Network Journal* 21(2019): 479–90.

Scarpello, Giovanni Mingari, and Daniele Ritelli. "Johann Bernoulli's First Lecture from the First Integral Calculus Textbook Ever Written: An Annotated Translation." *International Journal of Mathematical Education in Science and Technology* 50.6(2019): 839–55.

Scheepers, Christoph, et al. "Hierarchical Structure Priming from Mathematics to Two- and Three-Site Relative Clause Attachment." *Cognition* 189(2019): 155–66.

Schirn, Matthias. "Frege's Philosophy of Geometry." *Synthese* 196(2019): 929–71.

Schwartz, Richard Evan. "Pushing a Rectangle down a Path." *The Mathematical Intelligencer* 41.1(2019): 7–10.

Sepulcre, Juan Matías. "Public Recognition and Media Coverage of Mathematical Achievements." *Journal of Humanistic Mathematics* 9.2(2019).

Sereni, Andrea. "On the Philosophical Significance of Frege's Constraint." *Philosophia Mathematica* 27(2019): 244–75.

Shaw, Liam P., and Luke F. Shaw. "The Flying Bomb and the Actuary." *Significance* 16.5(2019): 13–17.

Sheldon, Neil. "What Does It All Mean?" *Significance* 16.4(2019): 15–16.

Siegelman, Noam, et al. "What Exactly Is Learned in Visual Statistical Learning? Insights from Bayesian Modeling." *Cognition* 192(2019): 1–10.

Simoson, Andrew "A Phyllotaxis of the Irrational." *Math Horizons* 26.3(2019): 10–13.

Skrodzki, Marti, Ulrike Bath, Kevin Guo, and Konrad Polthier. "A Leap Forward: A User Study on Gestural Geometry Exploration." *Journal of Mathematics and the Arts* 13.4(2019): 369–82.

Sokolowski, H. Moriah, Zachary Hawes, and Ian M. Lyons. "What Explains Sex Differences in Math Anxiety? A Closer Look at the Role of Spatial Processing." *Cognition* 182(2019): 193–212.

Soon, Low Chee. "A Sumptuous Buffet of Mathematical Strategies." *The Mathematics Teacher* 112.7(2019): 516–19.

Staats, Susan, and Lori Ann Lester. "About Time." *For the Learning of Mathematics* 39.1(2019): 44–47.

Sterpetti, Fabio. "Mathematical Knowledge and Naturalism." *Philosophia* 27(2019): 225–47.

Stewart, Ian. "Six Ages of Uncertainty." *Significance* 16.6(2019): 10–11.

Stoutenburg, Gregory. "In Defense of an Epistemic Probability Account of Luck." *Synthese* 196(2019): 5099–113.

Strobino, Riccardo. "Varieties of Demonstration in Alfarabi." *History and Philosophy of Logic* 40.1(2019): 42–62.

Tapp, Christian. "Bolzano's Concept of Divine Infinity." *The Infinity of God: New Perspectives in Theology and Philosophy,* edited by Benedikt Paul Gőcke and Christian Tapp. Notre Dame, IN: Notre Dame Press, 2019, 150–66.

Tatarchenko, Ksenia. "Thinking Algorithmically: From Cold War Computer Science to the Socialist Information Culture." *Historical Studies in the Natural Sciences* 49.2(2019): 194–225.

Teather, Anne, Andrew Chamberlain, and Mike Parker Pearson. "The Chalk Drums from Folkton and Lavant: Measuring Devices from the Time of Stonehenge." *British Journal for the History of Mathematics* 34.1(2019): 1–11.

Todorčević, Vesna, and Maria Šegan-Radonjić. "Mihailo Petrović Alas: Mathematician and Master Fisherman." *The Mathematical Intelligencer* 41.3(2019): 44–50.

Tou, Erik R. "Bernoullian Influences on Leonhard Euler's Early Fluid Mechanics." *British Journal for the History of Mathematics* 34.2(2019): 104–17.

Usó Doménech, J. L., and J. A. Nescolarde-Selva. "Mathematical Logic of Notions and Concepts." *Foundations of Science* 24(2019): 641–55.

Valdes, Juan E. Nápoles, and María N. Quevedo. "The Derivative Notion Revisited: The Fractional Case." *Mathematics Enthusiast* 16(2019): 369–76.

Vélez, Natalia, Sophie Bridgers, and Hyowon Gweon. "The Rare Preference Effect: Statistical Information Influences Social Affiliation Judgments." *Cognition* 192(2019): 1–11.

Venkat, Hamsa, et al. "Architecture of Mathematical Structure." *For the Learning of Mathematics* 39.1(2019): 13–17.

Verner, Igor, Khayriah Massarwe, and Daoud Bshouty. "Development of Competencies for Teaching Geometry through an Ethnomathematical Approach." *The Journal of Mathematical Behavior* 56(2019): 1–14.

Volkert, Klaus. "Note on Models." *Historia Mathematica* 48(2019): 87–95.

von Hippel, Matt. "The Particle Code." *Scientific American* 320.1(2019): 30–35.

von Mehren, Ann L. "Finding Teaching Inspiration from Gorgias: Mathematics Lessons from a Sophist." *Journal of Humanistic Mathematics* 9.1(2019).

Wagner, Roy, and Samuel Hunziker. "Jost Bürgi's Methods of Calculating Sines, and Possible Transmission from India." *Archive for History of Exact Sciences* 73(2019): 243–60.

Wainer, Howard. "A Statistician Reads the Obituaries: On the Relative Effects of Race and Crime on Employment." *Chance* 32.2(2019): 53–56.

Wapner, Leonard E. "Fair and Efficient by Chance." *Chance* 32.2(2019): 6–10.

Wares, Arsalan. "An Unexpected Property of Quadrilaterals." *International Journal of Mathematical Education in Science and Technology* 50.2(2019): 315–21.

Wilmott, Paul. "Smell the Glove." *Wilmott Magazine* 102(2019): 24–33.

Wilmott, Paul, and David Orrell. "No Laws, Only Toys." *Wilmott Magazine* 99(2019): 20–28.

Woll, Christian. "There Is a 3 × 3 Magic Square of Squares on the Moon—A Lot of Them, Actually." *The Mathematical Intelligencer* 41.2(2019): 73–76.

Yan, Xiaoheng, John Mason, and Gila Hanna. "Probing Interactions in Exploratory Teaching: A Case Study." *International Journal of Mathematical Education in Science and Technology* 50.2(2019): 244–59.

Yan, Yuan, and Marc G. Genton. "The Tukey *g*-and-*h* Distribution." *Significance* 16.3(2019): 12–13.

Yuan, Lei, et al. "Preschoolers and Multi-Digit Numbers: A Path to Mathematics through the Symbols Themselves." *Cognition* 189(2019): 89–104.

Zavoleas, Yannis, and Mark Taylor. "From Cartesian to Topological Geometry: Challenging Flatness in Architecture." *Nexus Network Journal* 21(2019): 5–18.

Zawislak, Stanislaw, and Jerzy Kopeć. "A Graph-Based Analysis of Anton Chekhov's *Uncle Vanya*." *Journal of Humanistic Mathematics* 9.2(2019).

Zelbo, Sian. "The Recreational Mathematics Activities of Ordinary Nineteenth Century Americans: A Case Study of Two Mathematics Puzzle Columns and Their Contributors." *British Journal for the History of Mathematics* 34.3(2019): 155–78.

Ziemba, Bill. "The Pick 6 and the Rainbow Pick 6." *Wilmott Magazine* 104(2019): 70–80.

Notable Journal Issues

"Astrostatistics." *Chance* 32.3(2019).

"Clinical Trials." *Chance* 32.4(2019).

"Affect and Mathematics in Young Children." *Educational Studies in Mathematics* 100.3(2019).

"Rituals and Explorations in Mathematical Teaching and Learning." *Educational Studies in Mathematics* 101.2(2019).

"Computational History and Philosophy of Science." *Isis* 110.3(2019).

"Explanation in the History of Science." *Isis* 110.2(2019).

"On Mathemata: Commenting on Greek and Arabic Mathematical Texts." *Historia Mathematica* 47(2019).

"The Roles and Uses of Examples in Conjecturing and Proving." *The Journal of Mathematical Behavior* 53(2019).

"Mathematics and the Arts in Education." *Journal of Mathematics and the Arts* 13.1–2(2019).

"Mathematical Misconceptions." *Mathematics Teacher* 112.6(2019).

"Mongeometry." *Nexus Network Journal* 21.1(2019).

"The Palladio Century." *Nexus Network Journal* 21.2(2019).

"The Emergence of Structuralism." *Philosophia Mathematica* 27.3(2019).

"The Creation and Implementation of Effective Homework Assignments." *PRIMUS* 29.1–2(2019).

"Mathematics for Social Justice." *PRIMUS* 29.3–4(2019).

"Modelling Approach to Teaching Differential Equations." *PRIMUS* 29.6–7(2019).

"Interdisciplinary Conversations." *PRIMUS* 29.8–9(2019).

"Forensic Science and Statistics." *Significance* 16.2(2019).

"Infinite Idealizations in Science." *Synthese* 196.5(2019).

"Between First- and Second-Order Logic." *Synthese* 196.7(2019).

"The Uses and Abuses of Mathematics in Early Modern Philosophy." *Synthese* 196.9(2019).

"Whole Number Arithmetic and Its Teaching and Learning." *Zentralblatt für Didaktik der Mathematik* 51.1(2019).

"Material Ecologies of Teaching and Learning." *Zentralblatt für Didaktik der Mathematik* 51.2 (2019).

"Identity in Mathematical Education." *Zentralblatt für Didaktik der Mathematik* 51.3(2019).

"Metacognition in Mathematics Education." *Zentralblatt für Didaktik der Mathematik* 51.4(2019).

"Mathematical Evidence and Argument." *Zentralblatt für Didaktik der Mathematik* 51.5(2019).

"21st Century Skills and STEM Teaching and Learning." *Zentralblatt für Didaktik der Mathematik* 51.6(2019).

"Research on Teaching and Learning in Linear Algebra." *Zentralblatt für Didaktik der Mathematik* 51.7(2019).

Acknowledgments

The pieces you can read in this volume were published in 2019, but we worked on preparing the anthology during the onset and apex (we hope!) of the coronavirus health crisis, over the spring of 2020. I thank all the contributors for writing the texts and for cooperating in the republishing in this form.

At Princeton University Press, thanks to Susannah Shoemaker for supporting me in this enterprise, to Kristen Hop for securing copyrights, to Nathan Carr for once again overseeing the production (eleventh time), and to Paula Bérard for copy editing.

Thanks to Leonid Kavalev and Jeffrey Meyer in the mathematics department at Syracuse University for assigning me a full load of courses and for scheduling my teaching conveniently. This has led to unprecedented income stability, allowing me to compound survival skills I developed in much more difficult times.

At home, thanks to my wife Fangfang, as well as to Ioana, Leo, and Ray for allowing me to teach remotely amid our busy tri-lingual household—and to work on this volume during early mornings and late nights!

Credits

"Outsmarting a Virus with Math" by Steven Strogatz. Previously published in *Scientific American*, 320.4(2019): 70–73. Also published in *Infinite Powers: How Calculus Reveals the Secrets of the Universe* by Steven Strogratz. Copyright © 2019 by Steven Strogatz. Reprinted by permission of Houghton Mifflin Harcourt and Atlantic Books Ltd. All rights reserved.

"Uncertainty" by Peter J. Denning and Ted G. Lewis. Originally published in *Communications of the ACM*, 62.12(2019): 26–28. Reprinted by permission of the authors.

"The Inescapable Casino" by Bruce M. Boghosian. Originally published in *Scientific American*, 321.5(2019): 70–77. Reproduced with permission. Copyright © 2019 SCIENTIFIC AMERICAN, a Division of Springer Nature America, Inc. All rights reserved.

"Resolving the Fuel Economy Singularity" by Stan Wagon. Originally published in *Math Horizons*, 26.1(2018): 5–9. Reprinted by permission of Mathematical Association of America.

"The Median Voter Theorem: Why Politicians Move to the Center" by Jørgen Veisdal. Originally published in *Medium*, October 11, 2019, https://medium.com/cantors-paradise/the-median-voter-theorem-c81630b57fa4. Reprinted by permission of Cantor's Paradise.

"The Math That Takes Newton into the Quantum World" by John Baez. Originally published in *Nautilus,* February 28, 2019. Reprinted by permission of Nautilus.

"Decades-Old Computer Science Conjecture Solved in Two Pages" by Erica Klarreich. Originally published in *Quanta Magazine* July 25, 2019. Original story reprinted with permission from Quanta Magazine (www.quantamagazine.org) an editorially independent publication of the Simons Foundation whose mission is to enhance public understanding of science by covering research developments and trends in mathematics and the physical and life sciences.

"The Three-Body Problem" by Richard Montgomery. Originally published in *Scientific American* 321.2(2019): 66–73. Reproduced with permission. Copyright © 2019 SCIENTIFIC AMERICAN, a Division of Springer Nature America, Inc. All rights reserved. Reprinted by permission of Springer Nature.

"The Intrigues and Delights of Kleinian and Quasi-Fuchsian Limit Sets" by Chris King. Originally published in *The Mathematical Intelliger*, 41.2(2019): 12–19. Reprinted by permission of Springer Science+Business Media, LLC, part of Springer Nature.

"Mathematical Treasures from Sid Sackson" by Jim Henle. Originally published in *The Mathematical Intelliger*, 41.1(2019): 71–77. Reprinted by permission of Springer Science+Business Media, LLC, part of Springer Nature.

"The Amazing Math Inside the Rubik's Cube" by Dave Linkletter. Orignally published in *Popular Mechanics Online,* December 16, 2019, https://www.popularmechanics.com/science/math/a30244043/solve-rubiks-cube/?utm_source=pocket-newtab. Reprinted with permission from Popular Mechanics.

"What Is a Hyperbolic 3-Manifold?" by Colin Adams. Originally published in *Notices of the American Mathematical Society* 65.5(May 2018): 544–46. Copyright © 2018 American Mathematical Society. Reprinted by permission.

"Higher Dimensional Geometries: What Are They Good For?" by Boris Odehnal. Originally published in *Proceedings of the 18th International Conference on Geometry and Graphics,* edited by Luigi Cocchiarella. Cham, Switzerland: Springer, 2019, pp. 94–105. Reprinted by permission of Springer International Publishing AG, part of Springer Nature.

"Who Mourns the Tenth Heegner Number?" by James Propp. Originally published in *Math Horizons* 27(2): 18–21. Reprinted by permission of the Mathematical Association of America.

"On Your Mark, Get Set, Multiply" by Patrick Honner. Originally published in *Quanta Magazine* September 23, 2019. Original story reprinted with permission from Quanta Magazine (www.quantamagazine.org) an editorially independent publication of the Simons Foundation whose mission is to enhance public understanding of science by covering research developments and trends in mathematics and the physical and life sciences.

"1994, the Year Calculus Was Born" by Ben Orlin. Excerpted from the book CHANGE IS THE ONLY CONSTANT by Ben Orlin. Copyright © 2019 by Ben Orlin. Reprinted by permission of Black Dog & Leventhal, an imprint of Perseus Books, LLC, a subsidiary of Hachette Book Group, Inc., New York, NY, USA. All rights reserved.

"Gauss's Computation of the Easter Date" by Donald Teets. Originally published in *Mathematics Magazine* 92.2(2019): 91–98. Reprinted by permission of the Mathematical Association of America.

"Mathematical Knowledge and Reality" by Paul Thagard. Adapted, with minor changes made by the author, from *Natural Philosophy: From Social Brains to Knowledge, Reality, Morality, and Beauty* by Paul Thagard. Oxford: Oxford University Press, 2019, pp. 272–86. Copyright © 2019 by Oxford University Press. Reproduced with permission of the Licensor through PLSclear.

"The Ins and Outs of Mathematical Explanation" by Mark Colyvan. Originally published in *The Mathematical Intelligencer* 40.4(2018): 26–29. Reprinted by permission of Springer Science+Business Media, LLC, part of Springer Nature.

"Statistical Intervals, Not Statistical Significance" by Gerald J. Hahn, Necip Doganaksoy, and William Q. Meeker. Originally published in *Significance* 16.4(2019): 20–22. © 2019 The Royal Statistical Society and American Statistical Association. Reprinted by permission.